12611505

WITHDRAWN
NDSU

ACTIO AND PERSUASION

ACTIO
AND
PERSUASION

Dramatic Performance in
Eighteenth-Century France

ANGELICA GOODDEN

CLARENDON PRESS · OXFORD
1986

Oxford University Press, Walton Street, Oxford OX2 6DP

*Oxford New York Toronto
Delhi Bombay Calcutta Madras Karachi
Kuala Lumpur Singapore Hong Kong Tokyo
Nairobi Dar es Salaam Cape Town
Melbourne Auckland*

*and associated companies in
Beirut Berlin Ibadan Nicosia*

Oxford is a trade mark of Oxford University Press

© *Angelica Goodden 1986*

*All rights reserved. No part of this publication may be reproduced,
stored in a retrieval system, or transmitted, in any form or by any means,
electronic, mechanical, photocopying, recording, or otherwise, without
the prior permission of Oxford University Press*

British Library Cataloguing in Publication Data

*Goodden, Angelica
Actio and persuasion: dramatic performance in
eighteenth-century France.
1. Theater—France—History—18th century
2. Acting—History
792'.028'0944   PN2633*

ISBN 0-19-815836-X

Library of Congress Cataloging in Publication Data

*Goodden, Angelica.
Actio and persuasion.
Bibliography: p.
Includes index.
1. Theatre—France—History—18th century
2. Acting—History
Intellectual life—18th century.   4. Rhetoric—1500–
1800.   5. Acting.   I. Title.
PN2633.G6   1986   792'.0944   85-21537*

ISBN 0-19-815836-X

*Set by Downdell Ltd., Oxford
Printed in Great Britain
at the University Printing House, Oxford
by David Stanford
Printer to the University*

Acknowledgements

In preparing this book I have enjoyed the facilities of many libraries and collections, and am particularly indebted to the following: in London, the British Library, the Victoria and Albert Museum, and the Warburg Institute; in Oxford, the Ashmolean Museum, the Bodleian Library, and the Taylor Institution; and in Paris, the Archives nationales, the Bibiothèque de l'Arsenal, the Bibliothèque historique de la Ville de Paris, the Bibliothèque nationale, and the Bibliothèque de l'Opéra. Dr Terence Cave, Professor Peter France, Mr Brendan McLaughlin, and Dr Denys Potts kindly read versions of the typescript and helped materially to improve it. The Faculty of Medieval and Modern Languages in the University of Oxford generously helped to finance periods of study in Paris, and St Hilda's College granted me an invaluable term of sabbatical leave in which to pursue research; I am very grateful to both of them. Finally, I owe thanks to the editors of *Forum for Modern Language Studies*, *French Studies*, and *Oxford Art Journal* for allowing me to quote from articles of mine which they originally published.

Contents

	List of Plates	viii
	Introduction	1
1	*Persuasion and the Visual Image*	28
2	*Action and Conviction*	48
3	*Acting and Visual Art*	73
4	*Pantomime Performances*	94
5	*'Bodily Eloquence' and Dance*	112
6	*Rules and Correspondences*	139
	Conclusion	161
	Select Bibliography	177
	Index	191

List of Plates

Between pages 40–41

1. Ziesenis, *Le 'Père de famille' de Diderot* (Paris, Bibliothèque nationale, Estampes)
2. After David, *Le Serment du Jeu de Paume* (Paris, Musée Carnavalet)
3. Moreau le Jeune, *Mirabeau arrive aux Champs Élysées* (Paris, Bibliothèque nationale, Estampes)
4. David, *Les Licteurs apportant à Brutus les corps de ses fils* (M. Ulysse Moussalli)

Between pages 136–137

5. David, *Les Sabines* (Paris, Musée du Louvre)
6. Carle Vanloo, *Mademoiselle Clairon en Médée* (Potsdam-Sanssouci, Bildergalerie)
7. Carle Vanloo, *Médée et Jason*, sketch (Pau, Musée des Beaux-Arts)
8. Watteau, *The Mountebank* (Oxford, Ashmolean Museum)

Introduction

In eighteenth-century France there was a determined attempt both to enhance the standing of actors and to establish the prestige of acting as a 'liberal' art—that is, one possessing an intellectual dimension beyond mere techniques or skills. Central to this attempt was a re-examination of the part played by bodily eloquence (physical attitude and gesture, and facial expression) in dramatic performance: such eloquence came to be regarded not as a superficial accompaniment to the playwright's text, but as a mode of rhetoric in its own right. This last was the title under which painting had won acceptance as a liberal art, for Renaissance theorists had drawn analogies between the orator's assembling and delivering of speeches and the artist's composition and execution of pictures. Later, comparisons with rhetoric lent the art of the theatre a similar prestige. By the end of the seventeenth century rhetorical theory had expanded in scope beyond forensic and pulpit oratory into literary and dramatic composition, taking with it the intellectual dignity of law and religion; and the ancient rhetorical category of *actio*, or bodily eloquence, provided a basis for conferring a like dignity on the actual performance of dramatic works. Just as its posited relationship with oratory had raised painting from the level of manual skill to liberal profession, so in the eighteenth century both painting and rhetorical theory, by emphasizing *depiction*, helped to demonstrate that acting was not just a means of conveying art, but an art itself.

Hence this book's title, and its starting-point. Factors other than rhetoric, however, contributed to the increasing attention paid to bodily eloquence in eighteenth-century France—the revival of the ancient genre of pantomime, the characteristic acting style of the Italian players in France, the example given by influential actors like Lekain. Dance, which likewise involved bodily eloquence, had been raised to a position of prestige by Louis XIV, and also played its part in the eighteenth-century debates: not only had it won the official approval which actors coveted for their own profession, but it contained a 'teachable' element which meant that it could legitimately be regarded as an art rather than a matter of improvisation.

The description of bodily eloquence in verbal terms capable of transcending the static, purely spatial quality of visual art inevitably raised problems. On a theoretical level, Descartes and the painter Lebrun had contributed to a classification of the physical attitudes associated with the passions (pathognomic ideas further developed at the beginning of the eighteenth century in the *Réflexions critiques sur la poésie et sur la peinture* of Dubos, whose discussion of poetry is mainly concerned with drama). In practice, however, teachers of acting found themselves faced with formidable difficulties when they looked for a method of recording performances so that they could be offered as a model to later generations of actors. Nor could the attempt to dignify an art of performance be helped by the persistence of the time-honoured view that plays were essentially vehicles for the word rather than for bodily enactment. Moreover, religious moralists, in the wake of Bossuet, continued to argue that the very fact of physical performance was likely to inflame the spectator's passions and so do harm to his immortal soul. It is only in the 1750s that the emphasis can be seen to have shifted, although the acting profession continued to have many detractors. From mid-century there was a concerted effort to demonstrate that acting was an occupation deserving the dignity of academic status, and that to continue denying its practitioners various social and civic privileges was a scandalous abuse. The Revolution finally accorded actors the rights they had long desired, but the struggle to obtain them had been protracted. The importance of this struggle, culminating in the actor's eventual admission to the pantheon inhabited by the practitioners of other liberal arts, and the approximation of his profession (in theory at least) to classically respectable models, are the main subject of this book.

One of the most devoted champions of bodily eloquence in dramatic performance was Diderot, although his own plays are so periphrastic and prolix that they can have offered actors very little scope for displaying their bodily skills. But however unsuitable Diderot's *drames* may have been as vehicles for *actio*, his dramatic theory implied its crucial importance as part of the playwright's overall pedagogic scheme. Moral persuasion was a principal objective of *Le Fils naturel* (1757) and *Le Père de famille* (1758), and for Diderot the actor's ability to manipulate his audience through gesture and movement was an essential means to that end (*see Plate 1*). It was Diderot's view, as stated by the character Dorval in the *Entretiens sur 'Le Fils naturel'*, that 'Nous parlons trop dans nos drames; et, conséquemment, nos acteurs n'y jouent pas assez.'[1] But other writings of his also bear witness to the preoccupations evident in his dramatic work. In the *Lettre sur les sourds et muets* (1751) Diderot mentions his practice as a theatre-goer of blocking his ears and watching only the mute play of the actors on stage, and remarks that very few performers emerge with credit from this test of their proficiency.[2] In *Le Neveu de Rameau* (first drafted in 1761) he describes at some length the Nephew's 'pantomime', or bodily action unaccompanied by words. For the impoverished and ill-endowed Nephew, mime often works as a replacement for reality. His art of gesture and impersonation allows him to conjure up the worldly possessions he would like to enjoy, to mimic the self-abasement before more powerful beings which he believes to be universal (and so to impress on the observer his conviction that everyone acts a part in the 'vile pantomime' of existence), and to transport himself and his audience to a rarefied sphere of artistic performance which circumstances prevent him from inhabiting. For both the Nephew and the onlooker, pantomime proves an economical means of making different realities present: it stimulates the imagination and emphasizes the idea that the world is full of people pretending to be other than they truly are. The Nephew's acting has the further advantage for him that it gives substance to what is illusory. In his impersonations he can pretend to conditions and states—of material wealth or of artistic superiority—which he does not in fact enjoy. Although an inferior musician, he longs to be an outstanding one, and his mimicry enables him to persuade both himself and others that he is truly accomplished.

It is often observed that Diderot's philosophical dialogues, which were not written to be performed on stage, are far more dramatic than the plays he wrote in the 1750s. The verbal eloquence of the former is more vividly realistic, and the situations they present are more arresting, than the rhetoric and action of the *drames*. Several of the dialogues, such as *Le Neveu de Rameau*, the *Supplément au Voyage de Bougainville*, and the *Paradoxe sur le comédien*, have indeed been successfully produced on stage. As a playwright, of course, Diderot had

---

[1] *Entretiens sur 'Le Fils naturel'*, in *Œuvres esthétiques* (henceforth *Œ*), ed. Paul Vernière (Paris, 1968), p. 100. Subsequent references to the *Entretiens* are to this edition.
[2] *Lettre sur les sourds et muets*, ed. Paul Hugo Meyer, in *Diderot Studies*, VII (1965), 52. Subsequent references to the *Lettre sur les sourds et muets* are to this edition.

necessarily to engage with a well-established dramatic tradition, and the weight of precedent may be a partial explanation for the conventional nature of *Le Fils naturel* and *Le Père de famille*. It is known that Diderot blamed the failure of *Le Fils naturel* when it was performed at the Comédie-Française on the unwillingness or inability of the Comédiens to adapt their habitual stiff and declamatory acting style to his new requirements. Certainly, his remarks in the *Lettre sur les sourds et muets* suggest that no tradition for this kind of acting existed; and the observation is indeed valid as far as the non-comic theatre in France was concerned.

Although the call for change contained in the *Entretiens sur 'Le Fils naturel'* was influential, the implications of Diderot's theoretical writings were not unchallenged. Some eighteenth-century critics questioned whether the visually perceived could ever exercise the force that verbal discourse possessed, and objected to such enterprises as Diderot's on the grounds that what is seen constricts the imagination rather than liberates it. Louis-Sébastien Mercier, for instance, argued that the highest value belonged to art which required men to exercise their interpretative faculty, and that the visible presented itself too unequivocally for such interpretation to be possible.[3] Others, conversely, reasoned that the art of gesture and movement, divorced from the word, made just such interpretative demands on the spectator's imagination. But Mercier's view gained in weight from the fact that it echoed a classical authority. According to the Platonic tradition the senses, whose objects were taken to be fleeting and unstable, could not be the source of true knowledge; only the intellect could reach the changeless world of the real. Lest it should appear that Plato's theory was largely irrelevant to an age which, like the eighteenth century, emphasized the role of sense-perception in the acquisition of knowledge, it is worth observing that Diderot at times subscribed to it himself. In the *Lettre sur les sourds et muets*, for example, he told the story of an imaginary people divided into five sects, each of which corresponded to one of the senses. When the sects quarrelled with one another, and mutual recrimination ensued, the sect associated with sight was condemned to the madhouse as visionary; that with smell, deemed imbecile; that with taste, adjudged intolerable because of its caprice and false delicacy; that with hearing, detested for its curiosity and pride; and that with touch, berated for its

[3] See some contributions of his to the *Journal de Paris* in 1797 (14 pluviôse an V/2 February 1797 *et seq.*); also below, chapter 1.

Introduction

materialism (p. 94). Although La Fontaine and La Motte, Diderot writes, could have made an excellent fable of this story, they would not have bettered Plato's allegory in the *Republic*, where sensory perception is shown as incapable of leading men to an understanding of the ultimately real.

The description in the *Republic* of prisoners in a cave whose perception of reality is confined to the observation of shadows on the wall clearly appealed to Diderot, for he returned to it in the *Salon* of 1765 in his commentary on Fragonard's painting *Corésus et Callirhoé*.[4] His review takes the form of the description of a dream following a day spent at the Salon, where he saw Fragonard's picture, and an evening during which he read some of Plato's dialogues. In the dream he imagines himself a prisoner in Plato's cave amidst men, women, and children in chains, who are fettered in such a way that they cannot turn their heads to see the world outside. Diderot observes that most of the prisoners seem to be happy in their shackles, occupying themselves with drinking, singing, and laughing, and that the few who try to escape from enslavement are frowned on by the others. All the prisoners, he writes, have their backs to the entrance to the cave and face the opposite wall, which is covered with a great canvas. Between them and the entrance stand a group of charlatans displaying a variety of modelled figures: shadows of these objects are thrown on to the canvas and form scenes so natural and true that the onlookers take them for real, sometimes laughing and sometimes weeping at what they see. The illusion is intensified by the activities of some rogues in the pay of the charlatans who stand behind the canvas and give the shadows the speech and the accent appropriate to the roles they are performing. Some of the assembled crowd are suspicious of what they see, shake their chains, and try to turn their heads; but the charlatans quell their protests with threatening cries. Diderot breaks off at this point, and moves to a description of the images formed by the shadows on the canvas (which prove to correspond to the objects in Fragonard's painting). But he promises to pursue his storytelling another time, and to relate to his interlocutor Grimm further consequences of the human enslavement to mere appearances of reality, the risks which the few rebels in the cave ran, and the persecution visited on those who showed their impatience at being subjected to illusion.

Despite employing such allegories as these, Diderot more often

[4] *Salons*, ed. Jean Seznec and Jean Adhémar, 4 vols. (Oxford, 1957–67), II. 189 ff. Subsequent references to the *Salons* are to this edition.

asserted the efficacy of sense-impressions in revealing reality than he denied it. In this he was at one with his age, impregnated as it was with Locke's doctrine of sensationalism. And for most sensationalists, as for many thinkers from antiquity onwards, the most powerful of the senses was sight: it gave man an access to reality more immediate than that furnished by the other four.

This emphasis helps to explain the origin of the ideas about acting which Diderot develops in the *Entretiens sur 'Le Fils naturel'*. But his ideas were shaped by other influences too. The art of pantomime, where the bodily eloquence of the actor took precedence over verbal discourse, enjoyed new favour in eighteenth-century France. Diderot's remarks on pantomime in the *Lettre sur les sourds et muets* echoed what Dubos had written in his *Réflexions critiques sur la poésie et sur la peinture* (1719), which contained a call for the revival of this ancient art. A later edition of Dubos's work, that of 1732, referred to private performances in the duchesse du Maine's château at Sceaux, where in 1714 the last act of Corneille's *Horace* was performed as a 'ballet d'action pantomime' modelled on the antique.[5] (The ballet-master Noverre later produced a balletic version of the story.) This edition also mentioned English versions of pantomime. In the early 1740s the Englishman Mainbray, an author of pantomimes and a ballet-master, staged very successful spectacles at the foire Saint-Germain in Paris.[6] Mainbray thus continued a tradition established a quarter of a century previously by the ballet-master Weaver, who produced both comic pantomimes and serious mythological ones. The 1740s saw a proliferation of such performances in Paris, not just at the fairs, but also at the Opéra, Théâtre-Italien, and Comédie-Française. François Riccoboni organized pantomime ballets at the Théâtre-Italien which were thought by contemporaries to signal a rebirth of the ancient art of *saltatio* (literally, jumping) and the pantomimic element in Greek drama.

The conventions governing the performance of serious drama in France were based partly on the rhetorical category of *pronuntiatio*, a term which could refer to bodily as well as verbal eloquence. Although these conventions excluded the degree of silent action present in comedy, the expressive style sought by actors like Lekain was influenced by contemporary interest in ancient pantomime. Voltaire, who

[5] See Kirsten Gram Holmström, *Monodrama, Attitudes, Tableaux Vivants* (Stockholm, 1967), p. 246, note 9.
[6] See Gösta M. Bergman, 'La Grande Mode des pantomimes à Paris vers 1740 et les spectacles d'optique de Servandoni', *Recherches théâtrales*, 2 (1960), 71–81.

organized his own private theatricals, was an enthusiast for this mode, and approved Lekain's striking of eloquent bodily postures in the performance of his tragedy *Sémiramis*. In his memoirs the actor Préville describes the 'tableau mouvant' which featured in a revival of this play, with Lekain (in the role of Arsace) frozen into immobility as he left Ninus's tomb.[7] It would be misleading, however, to suggest that actors and actresses in the first fifty or sixty years of the eighteenth century habitually performed tragedy in pantomimic style. Grand impressiveness remained a guiding principle in serious drama, and bodily eloquence was more commonly expressed in tableauesque poses than in the sequences of swift action characteristic of comedy.

Comic acting was very different from tragic, although many performers, like those of the Comédie-Française, were required to be proficient in both types. One reason for the favour extended to pantomime in eighteenth-century France was certainly the influence of the *Commedia dell'arte*, whose performing style involved a high degree of physical action.[8] The graphic images of Watteau, his master Gillot, and other eighteenth-century artists powerfully convey the gestural quality of Italian acting. The Italian troupe in Paris did not start acting in French until 1668, but before that their use of gesture meant that even spectators who did not understand Italian could follow the action of a play. The fact that masks were worn by the actors further necessitated the use of bodily eloquence, since their facial expression was fixed. In the mid-eighteenth century William Hogarth discussed the different movements of the *Commedia* types in terms which reveal their conventionalized nature, and the typical variations from one character to another. The action of the Harlequin, he writes in *The Analysis of Beauty*, is ingeniously composed of short, quick movements of the head, hands, and feet, 'some of which shoot out as it were from the body in straight lines, or are twirled about in little circles.' Scaramouche is gravely absurd as the character is intended to be, in overstretched, tedious motions made up of unnatural lengths of line. These two characters seem to have been contrived by conceiving a direct opposition of movements. The Pierrot's gestures and attitudes are chiefly in perpendiculars and parallels, as is the pattern of his dress; and Pulchinello is droll by being the reverse of all elegance, with

---

[7] *Mémoires de Préville et de Dazincourt*, ed. M. Ourry (Paris, 1823), p. 191.

[8] See Gustave Attinger, *L'Esprit de la Commedia dell'arte dans le théâtre français* (Paris, 1950), *passim*, and S. S. B. Taylor, 'Le Geste chez les "maîtres" italiens de Molière', *XVIIe Siècle*, 132 (1981), 285–301.

respect both to movement and to figure. The beauty of variety, according to Hogarth, is totally and comically excluded from this character: his limbs are raised and let fall almost simultaneously, 'as if his seeming fewer joints than ordinary were no better than the hinges of a door.'[9] Later in the *Analysis* Hogarth notes a mischief that attends copied actions on stage. They are often confined to certain sets and numbers which grow stale with repetition to the audience, and finally become subject to mimicry and ridicule. This would hardly be the case, he observes, if an actor were possessed of such general principles as include a knowledge of the effects of all movements the body can make (p. 153). Whether the typical Italian actor had such theoretical knowledge may be doubted, but the fact that spontaneity was a fundamental principle of acting style for the *Commedia dell'arte* suggests that performance remained varied enough to maintain an audience's interest over long periods of time.

Between the expulsion of the Italian actors in 1697 (resulting from an imagined slight to Mme de Maintenon in one of their plays) and their recall in 1716, much of their repertoire, and with it their acting techniques, had been adopted by performers at the fairs. But the success of theatricals at the foire Saint-Laurent and the foire Saint-Germain in eighteenth-century Paris also led to a development of non-verbal acting for another reason than simply the forains' assumption of the Italian style. The official troupes of the Comédie-Française, Comédie-Italienne, and Opéra saw fairground artistes as infringing the exclusive 'privilèges' for performing drama and opera which they had been accorded by royal decree in the seventeenth century, and therefore attempted to deprive the forains altogether of the right to act, dance, and sing. A consequence of this was the latter's resort to staging dramas without dialogue, in the performance of which they could be seen to obey the letter, if not the spirit, of the ordinances issued against them. It would be perverse, however, to argue that such prohibitions alone encouraged bodily acting in the unofficial theatres. The nature of their clientele—originally the uneducated or barely educated, although they gradually came to attract a more refined audience as well—itself made an emphasis on non-verbal entertainment natural and expedient.

Dubos's *Réflexions* also raised another matter which unquestionably influenced performing style, and especially 'geste', in the eighteenth-century theatre. He remarked that the passions all had their reflection

---

[9] William Hogarth, *The Analysis of Beauty* (London, 1753), p. 149.

in outward action and expression, and that 'bodily communication' was therefore capable of conveying the finest shades of meaning to the observer. The opinion that inward states may be physically expressed in an individual's attitude, action, and physiognomy was not original at the time Dubos wrote the *Réflexions*, however, for pathognomic theories of this kind reach back to antiquity. Descartes had developed his own mechanistic theory of emotional expression in the second half of the seventeenth century. His belief that the passions were controlled by the brain led him to lay particular emphasis on the way men's faces indicate their emotions, but he considered external movements in general to result from internal passionate states. The painter Lebrun carried over this theory into the visual arts, and at the end of the seventeenth century he lectured at the Académie royale de peinture et de sculpture on bodily, and particularly facial, expression of emotion. From the outset he stressed the psycho-physical mechanism which produces passionate expression. This means of conveying inward states externally was to be applied by painters as a way of overcoming the limitations of their non-temporal and non-narrative artistic medium, whose nature made the communication of certain types of feeling difficult. Their efforts were not, it must be said, always kindly received by critics, who thought that the artistic translation of emotion was often obscure or mishandled. As late as 1824 Stendhal complained in his review of that year's Salon that the school of the painter David 'ne peut peindre que les corps; elle est décidément inhabile à peindre les âmes.'[10] Yet despite Stendhal's comment it is undeniable that the Davidians did strive to become painters of the soul,[11] although it is equally evident that from the last two decades of the eighteenth century onward they produced many canvases filled with overstated facial expression and gestures. The fact that actors in eighteenth-century France were often encouraged to take paintings as models for their performance meant that the doctrines associated with Lebrun found their way into the theory and practice of acting as well.

Dubos himself extended the doctrines of Descartes and Lebrun by arguing that the physical expression of emotion might be almost without limit in its variety. Differences in age, sex, temperament, nationality, and social position, he wrote, lead inevitably to modifi-

---

[10] See George Levitine, 'The Influence of Lavater and Girodet's *Expression des sentiments de l'âme*', *The Art Bulletin*, XXXVI (1959), 33.

[11] See Jean Locquin, *La Peinture d'histoire en France de 1747 à 1785* (Paris, 1912), p. 80 and p. 162.

cations in the expression of feeling, and these differences cannot be communicated by words. In the view of Dubos, whose *Réflexions* are concerned with poetry primarily in its dramatic guise, only the silent action of the stage performer is capable of doing so.[12]

Another factor contributed to the stress which many theorists and some practitioners of drama laid on bodily eloquence, and it was quite unconnected with philosophical theories of expression, or with the 'popular' theatre of the Italian actors, fairground players, and other such performers of 'geste'. The question of the actor's position in society, which was much discussed in the eighteenth century, led writers to draw parallels between acting and the liberal art of rhetoric (commonly regarded as the fountain-head of all other arts)[13] in an effort to prove the former's right to an elevated status in the hierarchy of arts. In various respects the actor was clearly not comparable with the orator of antiquity, particularly in that he was not usually the author of the discourse he delivered. But he did share with the orator a concern with enhancing speech through bodily attitude, gesture, and facial expression. This part of eloquence was what the ancient rhetoricians had called *actio*, and according to them it was one of the most important constitutive elements of the orator's art. But, paradoxically, it was little discussed in classical rhetorics. In seventeenth- and eighteenth-century works on the art of the orator's successors, the preacher and the advocate, the habitual failure to elucidate this aspect of eloquence was often explained in terms of the difficulty of describing physical movement and attitude in words. The same difficulty was referred to by many commentators on acting. None the less, as we have seen, it was frequently assumed that visual impressions have a more immediate effect than impressions conveyed by the other senses, and that in view of their importance they call for some sort of verbal translation. Many regarded the human impressionability to visual signals as legitimizing the use of *actio* in oratory, although the need for caution and control in the pulpit and courtroom was recurrently emphasized too. The goal of the orator was taken to be persuasion, and it was axiomatic for writers that the desired end could be more readily achieved when speech was supplemented by

---

[12] Jean-Baptiste Dubos, *Réflexions critiques sur la poésie et sur la peinture*, 2 vols. (Paris, 1719), I. 514–15.

[13] Cicero's references to poetry, music, painting, architecture, sculpture, and the art of the actor often imply his belief that eloquence is the regulating principle of all the other arts. See Marc Fumaroli, *L'Âge de l'éloquence* (Geneva, 1980), p. 51.

bodily movements and gestures than when verbal means alone were employed. In seventeenth- and eighteenth-century treatises, it is true, the actor was often unfavourably compared with the more dignified preacher or advocate. But the notion that his bodily action was based on the same principles as theirs, which themselves derived from the *actio* of the ancient orator, was seen by some as suggesting that the analogy between acting and better-accredited professions was legitimate.

Theorists of other arts had tried to show how those arts too were related to rhetoric, and writers on acting drew on their arguments. The most noteworthy example had been provided by theorists of painting in Italy, who claimed that painters followed all the procedures, and shared all the persuasive intentions, of orators preparing and delivering a speech. According to Alberti's *Della pittura* (1436) the aim of the painter, like that of the classical orator, was to please, move, and convince.[14] In the humanist ambience of fifteenth-century Florence it was natural for theorists to turn to rhetoric in their effort to find for painting a model that would provide it with 'classical' respectability. In *De oratore* Cicero had likened one part of the orator's (and actor's) equipment with the painter's, stating that the voice of the former functioned like the colours used by the latter in rendering nuances of emotion (III. lvii. 216–17). Alberti proceeded to give instructions on the use of colours in painting that were appropriate to the arousing of particular emotions in the observer. The painter, it was argued by Renaissance writers, organized the parts of his composition according to the same principles as those observed by the orator in constructing his speech. A high proportion of ancient rhetorical terms were used as metaphors for visual experience. The words *gracilis* and *vehemens*, which were applied to paintings, referred back to one of the orator's three *genera dicendi*, or levels of style.[15] In Alberti's discussion of *compositio*, a model derived from rhetoric[16] was transferred to visual art. In the former, *compositio* signified the assembling of a sentence or period: the accepted hierarchy had it that words

[14] See John R. Spencer, 'Ut rhetorica pictura', *Journal of the Warburg and Courtauld Institutes*, 20 (1957), 26–44. The book by Franciscus Junius, *De pictura veterum* (1637), pointed to countless connections between ancient rhetoric and modern art, and was familiar to Poussin, Fréart de Chambray, Du Fresnoy, Félibien, and de Piles. Junius draws particularly on Cicero in his discussion of pictorial art in terms of the ancient divisions of rhetoric. See Colette Nativel, 'Franciscus Junius et le *De pictura veterum*', *XVIIe Siècle*, 138 (1983), 7–30.

[15] See Michael Baxandall, *Giotto and the Orators* (Oxford, 1971), pp. 17–18.

[16] See Aldo Scaglione, *The Classical Theory of Composition* (Chapel Hill, 1972).

make up phrases, phrases clauses, and clauses sentences. Similarly in painting, it was argued, planes make up members, members bodies, and bodies the coherent scenes of narrative art (Baxandall, p. 131). Furthermore, theorists stated that in the gestures and attitudes of his human figures the artist built on what ancient rhetorics had suggested concerning the types of movement the orator should use to express particular emotions and arouse them in his audience. From Cicero's brief indications Alberti constructed a science of gesture which introduced into the art of painting the notion that bodily movements reflect inward emotions. In late seventeenth-century France these notions were expanded into the doctrine of passionate expression associated with Lebrun, and subsequently were applied to the actor's art.

Alberti's views accorded with the new status of the painter as a practitioner of a liberal art, and led him to stress the need for painters to be educated in literary, historical, and scientific subjects. Their activity was acknowledged to be a dignified exercise of the spirit rather than a lowly manual skill. Leonardo protested in his *Trattato della pittura* (first published in 1651) against the depreciation of art implicit in its classification as a merely technical occupation, drawing attention to its scientific and intellectual nature; and this view of art led to the rational, academic conception of the subject that was to prevail for centuries to come. Although it was to be challenged towards the end of the eighteenth century, it exercised a force over general opinions concerning the status of painting which actors were later to envy, and which led them, particularly from the 1750s, to call for a similar academic status for their profession. Thus actors, along with theorists of the subject, compared stage performance with aspects of painting and sculpture as well as with oratory. Interestingly, however, the depiction of types of bodily attitude or expression felt to be inappropriately reminiscent of visual art was sometimes criticized in eighteenth-century commentaries on stage performance, just as 'theatricality' of style was sometimes condemned in painting (as in ancient and modern treatises on oratory).

In the first half of the eighteenth century the artist and dramatist Charles-Antoine Coypel[17] tried to introduce theatrical elements into the existing academic theories of expression in painting,[18] although

---

[17] See I. Jamieson, *C.-A. Coypel, Premier Peintre de Louis XV et auteur dramatique. Sa Vie et son œuvre artistique et littéraire* (Paris, 1930).

[18] See F. Ingersoll-Smouse, 'Charles-Antoine Coypel', *Revue de l'art ancien et moderne*, XXXVII (1920), 143–54, 285–92.

his efforts were often disliked by contemporaries. The connoisseur and collector Mariette observed disapprovingly that Coypel's father Antoine, a past director of the Académie royale de peinture et de sculpture, had overemphasized the expressions of characters in his own work. In 1721 the elder Coypel delivered a speech at the Académie in which he declared that the artist needed an acquaintance with the rules of declamation in order to match gesture with facial expression, but that as his art was a non-verbal one he should make use of attitude and gesture to express meaning, as the dumb were obliged to do.[19] Anticipating Mariette's criticism, however, he warned against the temptation to employ extreme gesture resorted to by 'des esprits emportés et déréglés', and who abandoned the pursuit of truth, nature, and reason (ibid., p. 162). According to Mariette, Charles-Antoine Coypel also sought models for attitude and expression in the theatre, but seemed only to find the depiction of excess there: his canvases teemed with exaggerated attitudes, grimaces, and other facial expressions which utterly failed to convey sentiments of the soul.[20] Coypel abandoned his dramatist's career with regret, but as a painter he maintained his contacts with the theatre, producing a number of pictures on dramatic subjects for the Gobelins factory to translate into tapestries.

The younger Coypel saw the theatre as the best school for painters desirous of learning the art of intense expression, and wanted to carry the means of conveying passion on canvas further than Lebrun had done. He eschewed the depiction of momentary expression for the attempt to sum up in attitude and countenance an entire character, and even an entire play. In a lecture delivered at the Académie in 1749, 'Réflexions sur l'art de peindre en le comparant à l'art de bien dire', he discussed some dramatic moments from Racine's *Andromaque*, and remarked in the spirit of his father that

Les acteurs que nous mettons en scène n'ont d'autre langage que le geste et les mouvements du visage: en parlant il n'est point d'homme qui ne puisse aisément faire comprendre à quel point il est combattu par deux sentiments contraires, mais ce serait le chef-d'œuvre d'un muet que de pouvoir, en pareil cas, nous mettre au fait des mouvements opposés qui l'agitent.[21]

[19] *Procès-verbaux de l'Académie royale de peinture et de sculpture*, ed. A. de Montaiglon, 10 vols. (Paris, 1875-92), IV. 188.
[20] P. J. Mariette, *Abecedario*, ed. R. de Chennevières and A. de Montaiglon, 6 vols. (Paris, 1851-60), II. 32.
[21] See Antoine Schnapper, '"Le Chef-d'œuvre d'un muet", ou La Tentative de Charles Coypel', *Revue du Louvre et des musées de France*, XVIII (1968), 262. Other

Such advice as that offered by the two Coypels to students of painting implies a reversal of the order of priority between painting and performed drama which usually obtained in eighteenth-century France. For contemporaries, the point of comparing acting with painting was more often to show how the former deserved the prestige which the latter enjoyed than to suggest that painting ought to model itself on the theatre. But there were exceptions to that general rule.

In England, of course, the respectability of actors came earlier than in France. (Voltaire observed in the *Lettres philosophiques* that the English actress Sara Oldfield was buried alongside Newton in Westminster Abbey, whereas the French actress Adrienne Lecouvreur had been refused the simple dignity of Christian burial within the city walls of Paris.)[22] This respectability seems to have encouraged English writers to draw comparisons like those drawn by the Coypels between the theatre and the painter's canvas. In the early nineteenth century it was possible for Gilbert Austin, the author of a handbook on recording the actor's movements, to reverse the order of precedence between stage and picture which eighteenth-century commentators on acting had customarily maintained. Austin suggested that a principal advantage of the notational scheme he had developed, applied to acting, would be its ability to preserve records of stage performance for the use of history painters. A scene from Shakespeare, notated after the action of a great actor, would prove an object of incessant and fruitful study for the artist.[23] Austin's following words seem appropriate to what the artist David had in fact done during the French Revolution in his painting of the *Serment du Jeu de Paume*, although he mentions no painters by name:

> If the historical painter would hand down to posterity truth instead of fiction, and so stamp a double value on the productions of his pencil, he would represent the actual manner of gesture of the speakers concerned in the great events which he celebrates . . . when the art of recording gesture and expression shall have arrived at the perfection of which it is capable, the record of the manner will be as capable of being truly represented as the record of the words. (Ibid.)

commentators referred to the supreme difficulty of depicting composite passions: see Levitine, p. 40.

[22] Voltaire, *Lettres philosophiques*, ed. F. A. Taylor, 2nd edition (Oxford, 1946), p. 87; also Jean-Jacques Olivier, *Voltaire et les comédiens interprètes de son théâtre* (Paris, 1899), p. xix ff.

[23] Gilbert Austin, *Chironomia, or A Treatise on Rhetorical Delivery*, ed. Mary Margaret Robb and Lester Thonssen (Carbondale and Edwardsville, 1966), pp. 285–6.

Among paintings which borrowed from a theatrical model, David's *Brutus* (1789) was clearly based on a scene from Voltaire's tragedy of the same name; and other works of his, like *Le Serment des Horaces* and *Socrate au moment de boire la ciguë*, were influenced by dramas too.

Another suggested source for *Le Serment des Horaces* is Noverre's action ballet *Les Horaces et les Curiaces,* and dance, like painting, was often mentioned in discussions of bodily eloquence. Theorists of acting were aware that, here too, a consideration denied over much of the eighteenth century to the actor had been conferred on practitioners of a different performing art in virtue of the academic status of their profession. An academy for dancing had been founded in the seventeenth century, and Louis XIV had decreed that no one of noble birth should be deemed to have forfeited his status through adopting this career. Some took the view that acting was essentially different from dancing in that the former, unlike the latter, could not be reduced to rule. While it could be reasoned—as did the actress Mlle Clairon—that this claim in itself implied the superiority of acting to dancing, for others the difficulty of teaching actors by precept, as the dancer's and the painter's art could allegedly be taught, meant that acting might never have an academic status. At the end of 1772, it is true, the actor Préville obtained a 'privilège' for the founding of such an academy, which probably opened in 1773; but some critics argued that Préville's attention ought to be exclusively concentrated on the Comédie-Française (of which he was a member), and other actors in the troupe accused him of trying to establish a second company to compete with their own.[24] Mlle Contat seems to have been the only noteworthy product of this school, which was closed when the War of American Independence proved too great a drain on royal finances for inessential expenses to be continued. Subsequently the art of music came to the rescue of acting. An establishment independent of the Académie royale de musique, or Opéra, was opened to train actors for the latter, and the pensionnaires of the Comédie-Française were named as founding pupils. The Menus Plaisirs du Roi, which officially supervised royal entertainment, was thereby enlarged, and in 1786 the duc de Duras requested the institution of an École de Déclamation.[25] The actors Molé, Dugazon, and Fleury all

[24] See Constant Pierre, *Les Anciennes Écoles de déclamation dramatique* (Paris, 1896), pp. 9-13.
[25] See J.-G. Prod'homme and E. de Crauzat, *Les Menus Plaisirs du Roi, l'École royale et le Conservatoire de musique* (Paris, 1929), p. 101. The duc de Duras was one of the Premiers Gentilshommes de la Chambre du Roi, the official overseers of the King's entertainment.

worked for this academy. (In the absence of public schools of acting, approved actors had previously given private lessons to the aspiring young.) In 1789, however, it was closed, probably because of disagreement in high places about the need for such an establishment.

Teachers of acting often expressed their frustration at the difficulties that faced them. A common complaint, significantly, was that no satisfactory record existed of past performance on stage which could be offered as a model to the student. Here, according to some commentators, acting clearly diverged from dance, where techniques of choreography both permitted precept and facilitated record. (Not everyone, it should be said, agreed that choreography was of much use in the latter respect.) The German Engel made an effort towards the end of the eighteenth century to provide something resembling these methods for acting, but his success was comparatively slight.[26] There seems to have been no attempt in France to codify acting movements comparable to Engel's or, later, Austin's, although many French theorists discussed this matter and regretted the ephemeral nature of the actor's performance, and despite the fact that choreography had a predominantly French pedigree.

Further prejudice to the cause of bodily acting arose from the time-honoured view that plays were essentially vehicles for the word rather than for action. This opinion, whose influences may be felt in the 'closet' dramas of the sixteenth century as well as in the emphasis on declamation of the seventeenth, derived from Aristotle. In the *Poetics* he declares that well-written plays do not require performance for their vivid quality (*enargeia*) to be realized. Reading aloud suffices for that, although enactment will always increase the impression a drama makes (*Poetics*, §26).

Early in the eighteenth century Levesque de la Ravaillère discussed this belief, developing the notion in a manner which recalled a number of comparisons drawn in books of rhetoric between written and enacted eloquence:

> une déclamation sublime, expressive et animée contribue autant que la poésie aux plaisirs et à la perfection du théâtre: il semble que le secret de remuer le cœur, qui est l'unique objet du poème dramatique, lui soit destiné: elle seule sait le surprendre et l'émouvoir, lorsqu'il est presque indifférent et comme immobile, plus puissante que la poésie qui n'aurait pu que l'ébranler, elle l'entraîne et elle le remplit de passion.[27]

[26] J.-J. Engel, *Ideen zu einer Mimik*, 2 vols. (Berlin, 1785-6).
[27] [Pierre-Alexandre Levesque de la Ravaillère,] *Essai de comparaison entre la déclamation et la poésie dramatique* (Paris, 1729), p. 9.

Introduction

What is of interest here is Levesque's belief that an injustice has been done to the actor in the assumption that dramatic poetry is primarily a written text, and only secondarily an enacted one. He remarks that poetry has traditionally won the laurels; the perfection of the actor who gives it embodiment has merely been politely applauded. But Voltaire's opinion is singled out as an exception to this rule, as Levesque quotes approvingly his lines, addressed to Mlle Duclos,

> Vous partagez entre Racine et vous
> De notre encens le tribut légitime.
> (p. 11)

Levesque's opinion that actors over the ages have been denied the admiration they merit is no doubt an exaggerated one. The seventeenth century in France provides more than one instance of the actor's being accorded general esteem and consideration, and the fact that Boileau devotes a section of his *Art poétique* to the performing art of acting (as Levesque notes) indicates an awareness at the time of the player's importance. None the less, the *Essai de comparaison* provides an interesting early example of beliefs that were to gather force as the century advanced.

Aristotle's view certainly damaged the cause of dramatic production. But another opinion brought to mind by Levesque's invocation of Voltaire (who in the *Lettres philosophiques* deplores the abject treatment of actors by moralists and others) was equally harmful. It is epitomized in what Bossuet has to say about drama in the *Maximes et réflexions sur la comédie* (1694).[28] According to Bossuet, whose view expressly counters that of contemporary apologists for the theatre, plays present a particular danger to public morals because of their living quality. He compares a dramatic performance with a painting of a subject offensive to conventional morality, and says that the former's effect on human passions is greater than the latter's. A static painting, for Bossuet, inflames the beholder only moderately, whereas the action of the stage performer insidiously arouses the strongest emotions.

My emphasis in this book is on the visual dimension of the actor's performance rather than on his verbal eloquence, and on the importance of *actio* both in his own profession and in other performing arts.

[28] A. Vulpian and Gautier, *Code des théâtres* (Paris, 1829), p. 183; Bossuet, *Maximes et réflexions sur la comédie*, 4th edition (Paris, 1930), p. 10; Jean Dubu, 'Bossuet et le théâtre: un silence de l'Évêque de Meaux', *Journées Bossuet: La Prédication au XVIIe siècle*, ed. Thérèse Goyet and Jean-Pierre Collinet (Paris, 1980), p. 201.

In the first chapter I consider prevalent beliefs in the age of Diderot about the immediacy which visual evidence confers on the perception of reality, and relate them to the case of the performer's *actio* in the theatre. I then turn in chapter 2 to an examination of the actor's bodily eloquence as compared with that of the ancient orators and their eighteenth-century counterparts. This and subsequent chapters also pursue the suggestion that statements about the interrelationship of the arts—a commonplace of eighteenth-century aesthetics—were based at least partly on a desire to demonstrate the elevated status of particular arts like acting by positing their similarity to better-accredited ones. They thus played a more important part in the comparative evaluation of artistic production than the tired repetition of Horace's analogy in the *Ars poetica*, 'ut pictura poesis', might lead one to suppose.

The theory of such relationships had a long history. Aristotle as well as Horace had suggested correspondences between painting and poetry, and treatises on art and literature from the mid-sixteenth to the mid-eighteenth century constantly pointed to the close connections between them (often taking poetry in the sense of drama). An assumption frequently made was that they resembled one another in their common effort to imitate nature, but nature of a superior rather than an average kind. However, all Horace had meant to suggest in the *Ars poetica* by the phrase 'ut pictura poesis' was that both oratory and painting sometimes require minute examination, and sometimes need to be considered from afar without particular attention being paid to detail.[29] Later theorists took the matter further, and argued that all the arts, not merely those of painting and poetry, were fundamentally similar. Charles Batteux sought to demonstrate this resemblance in *Les Beaux-Arts réduits à un même principe* (1746), which again stated that the arts were united in their objective of imitating nature; but critics, like the Diderot of the *Lettre sur les sourds et muets*, complained that Batteux never explained what this nature was, so that his treatise added little to reasoned aesthetic discussion of the subject (*Lettre sur les sourds et muets*, p. 81). In the eighteenth century, in any case, there were dissenters from the view that visual and literary art are essentially the same. Shaftesbury, Dubos, Diderot, and Lessing, for example, reasoned that the former has no temporal dimension, but only a spatial one, whereas arts whose medium is language require

---

[29] See Wesley Trimpi, 'The Meaning of Horace's Ut pictura poesis', *Journal of the Warburg and Courtauld Institutes*, 36 (1973), 1-34.

narrative or expository time.[30] None the less, the theory of correspondence was an important part of aesthetic discussion during this period, and was often invoked in contemporary efforts to argue the dignity of acting.

The chapters on pantomime and dance, which both involved the performer's bodily movement and attitude, explore an important subsidiary reason why performing arts of this kind should have been popular in eighteenth-century France. Although there was a courtly tradition of dance, reflected in the academic status it enjoyed, it held an evident appeal for the common people in its less refined forms. Rousseau was not the only educated contemporary observer to remark on the intense communal pleasure to be derived from this activity. Dance, like the mute antics of the tumbler and acrobat at the fairground or the bodily exuberance of Harlequin at the Théâtre-Italien, spoke to all in virtue of its wordlessness, its emphasis on *actio* alone. In this it was theoretically akin to the mute art of painting. Dissenters from the idea that *pictura* and *poesis* were equivalent sometimes observed in this connection that painting (at least in its representational form) communicates directly, whereas the verbal medium of *poesis* is purely conventional, and cannot be immediately understood. Clearly, there are important respects in which painting does call for an interpretative effort on the part of the onlooker, but there is an element of truth in the notion that it conveys its meaning straightforwardly, and that its human configurations exercise an appeal related to that of the silent performer on stage or fairground trestle.

The potential appeal of spectacle (in a broad sense) to the unlettered also explains the popularity of another institution which underwent a significant development in late eighteenth-century France, and which at first sight seems unconnected with visual or performing art. The meetings of the national assembly after the events of 1789 enjoyed great favour among the populace as a form of entertainment, as well as being relished by serious political observers and cultivated *amateurs* of debate; and a part of this favour seems to have derived from the 'actorly' comportment of députés. Some, like the advocate and Revolutionary leader Hérault de Séchelles, were known to have taken

[30] On the connections between literary and visual arts see, *inter alia*, W. G. Howard, 'Ut pictura poesis', *PMLA*, 24 (1909), 40–123; Rensselaer W. Lee, 'Ut pictura poesis: The Humanistic Theory of Painting', *The Art Bulletin*, XXII (1940), 197–269; Rémy G. Saisselin, 'Ut pictura poesis: Du Bos to Diderot', *Journal of Aesthetics and Art Criticism*, XX (1961-2), 144–56; and E. H. Gombrich, 'Moment and Movement in Art', reprinted in his *The Image and the Eye* (Oxford, 1982), pp. 40–62.

lessons from performers on the professional stage, and the determination of politicians to impress the people as well as their peers often found expression in the cultivation of oratorical techniques based on those described by the ancient rhetoricians. Indeed, députés were also called 'orateurs'. Although the performance of speakers at the Assemblée and Convention depended above all on the verbal content of their discourse, there is a remarkable emphasis by many writers of the time on the visual dimension of their contributions, with the example of Mirabeau's acting skills the most prominent. Political commentators often referred, usually disparagingly, to the 'comédiens' charged with conducting the nation's business, and to the fact that the common people regarded their antics as an agreeable and free substitute for the theatre. While the *actio* of performers might have the dignity of classical precedent, it often seemed calculated to appeal above all to the 'petit peuple' whose involvement in political life many of the Revolutionaries sought to effect. As such, it was disapproved by conservative observers like the influential Swiss journalist Mallet du Pan, and the anti-Revolutionary Edmund Burke. This is not to say that all links between the worlds of politics and the theatre were regarded with amusement or contempt, or deserved to be. The actor Molé, for example, greatly admired Mirabeau, and regarded him as having a profound, not a trivial, feeling for dramatic performance. Equally, Talma's acquaintance with a number of the Girondins who frequented Julie Talma's salon was an eminently respectable one. But it is clear from the favour which the Assemblée's proceedings enjoyed among the barely educated classes that something akin to the rumbustious fun of popular theatre, with its emphasis on the seen as well as, or instead of, the heard, was sought and found there.

The accessibility to ordinary people of 'arts de spectacle' was deployed by the Revolutionary leaders in another institution which they turned to new ends after 1789. Fêtes had played their part in the courtly world of the ancien régime, but had rarely been directed at the populace. Their celebrations were of and for the monarchy and those who principally upheld it. But in Revolutionary France the fête became an important semi-dramatic production designed for the consumption of all, and this was reflected in the nature of the performances it embraced. In theory it exhibited the combination of all the arts—of music, the image, and the word—but commentators recurrently stressed the part played by pantomime and dance in its realization. Unsurprisingly, theorists of the fête often referred approvingly to the effect which such arts made on the spectator's senses rather than on his intellect, and thus set their observations in a tradition to which eighteenth-century ideas on *actio* belonged. Boissy d'Anglas's essay on national fêtes, for example, does not discount the importance of invoking man's reason as well as charming his senses as means to the necessary end of educating him politically and morally, but asserts that 'il faut parler à son âme et à son cœur non moins qu'à son esprit et qu'à sa raison, il faut éclairer et former l'un et l'autre'.[31] Many commentators seem to echo in their remarks what Rousseau had earlier written about spectacles arranged for the public at large: 'On ne saurait croire à quel point le cœur du peuple suit ses yeux, et combien la majesté du cérémonial lui impose.'[32] On the other hand, some writers criticized the Revolutionary fêtes for their concentration on the bodily action of participants, as though it were a degrading emphasis. Grobert, for instance, denied that fêtes should be simple assemblages of pantomimes, objecting that to focus on spectacle might be to subvert the underlying spiritual message of these celebrations.[33] Yet it was clear to their organizers that for the often abstract propagandist message to be effectively diffused among the people,[34] such visual elements were of prime importance.

This, together with the other factors discussed above, suggests several reasons why the actor's status should have been a subject for discussion in eighteenth-century France, particularly towards the end of the period. Since a principle of the Revolution was that all men are equal, there was particular significance in such mutual endeavours as the fête was intended to encourage, and the removal of barriers which separated the enacting parties from the passive audience. The need for interaction which Rousseau noted when he spoke against the imposition of obstacles between men and their fellows was argued by other writers too—the title of one of La Revellière-Lépeaux's works, significantly, is *Essai sur les moyens de faire participer l'universalité des spectateurs à tout ce qui se pratique dans les fêtes nationales* (an VI). In this work

---

[31] Boissy d'Anglas, *Essai sur les fêtes nationales, adressé à la Convention nationale* (Paris, 12 messidor an II/30 June 1794), p. 13.
[32] J.-J. Rousseau, *Considérations sur le gouvernement de Pologne*, in *Œuvres complètes*, ed. Bernard Gagnebin and Marcel Raymond, 4 vols. (Paris, 1959–69), III. 962–3.
[33] J. Grobert, *Des fêtes publiques chez les modernes* (Paris, an X), p. 34.
[34] See William Olander, 'French Painting and Politics in 1794: The Great *Concours de l'an II*', *Proceedings of the 10th Convention on Revolutionary Europe, 1750–1850* (1980), II. 21–2.

Lépeaux argues that the involvement of the people is essential to the success of the fête: 'ce n'est que pour le peuple, qui est spectateur, que sont instituées les fêtes publiques, et non pour le petit nombre de citoyens, quels qu'ils soient, qui y figurent.'[35] The notion that the people should be bystanders rather than participants, according to Lépeaux, is a misguided one: 'Quant aux cérémonies qui se pratiquent, je conviens que tous les assistants n'y participent pas d'une manière aussi immédiate; ils le font cependant jusqu'à un certain point' (p. 13). By a skilful involvement of the different sectors of the audience, as many as 2,000 or 3,000 citizens can simultaneously and actively participate, and 'éprouve[r] à la fois les mêmes impressions et partage[r] les mêmes jouissances. Je vais plus loin, je veux que pendant quelques instants tous ensemble ils soient acteurs eux-mêmes' (p. 14). Another work of Lépeaux's, the *Réflexions sur le culte, sur les cérémonies civiles et sur les fêtes nationales*, similarly advances the view that the citizens present at fêtes 'doivent être acteurs eux-mêmes, autant qu'il est possible'.[36] The example of the carnival comes to mind: the people ignore accepted divisions, both the division of one social class from another and that between actor and audience. Rousseau's distaste for the theatre is formulated in the *Lettre à d'Alembert* partly in terms of the separation of stage from spectator, for he contends that theatres are institutions 'qui renferment tristement un petit nombre de gens dans un antre obscur; qui les tiennent craintifs et immobiles dans le silence et l'inaction.'[37] Merlin de Thionville expressed

[35] L. M. La Revellière-Lépeaux, *Essai sur les moyens de faire participer l'universalité des spectateurs à tout ce qui se pratique dans les fêtes nationales* (Paris, an VI), p. 9.

[36] L. M. La Revellière-Lépeaux, *Réflexions sur le culte, sur les cérémonies civiles et sur les fêtes nationales*, in J.-F. Dubroca, *Discours sur divers sujets de morale et sur les fêtes nationales* (Paris, an VII), p. 41. For an earlier expression of such beliefs, see Du Pont's letter to Margrave Carl Friedrich of Baden on 31 December 1772: 'les spectacles du peuple sont des fêtes. Non pas des fêtes oisives . . ., mais des fêtes . . . dans lesquelles il puisse être acteur et non simplement spectateur. Car la fonction d'un spectateur immobile a quelque chose de lâche et d'amollissant, et devient souvent ennuyeuse; au lieu que ceux qui ont un rôle à remplir et des applaudissements à mériter et à recevoir ne s'ennuient jamais, et si leur rôle a vraiment de la noblesse et de la beauté, ils s'en pénètrent. Ce qui fait qu'il se forme actuellement en Europe tant de comédies bourgeoises, c'est qu'il en est en effet plus agréable de la jouer que de la voir: et ce qui fait que les femmes de tout pays préfèrent le bal à la comédie, c'est qu'il vaut mieux être regardé que regardant . . . L'homme est fait pour agir. Si nous voulons donc procurer au peuple des spectacles où ce tyran froid et cruel, ce corrupteur du monde, l'ennui, ne puisse pénétrer, que ce soit le peuple même qui les donne.' (*Carl Friedrichs von Baden brieflicher Verkehr mit Mirabeau und Du Pont*, edited by Carl Knies, 2 vols. (Heidelberg, 1892), II. 17–18.)

[37] J.-J. Rousseau, *Lettre à M. d'Alembert sur les spectacles*, ed. M. Fuchs (Lille and Geneva, 1948), p. 168.

the same sentiment in a remark about a planned ceremony addressed to the Convention nationale on 4 vendémiaire an II (25 September 1793): 'que dans cette fête le peuple n'ait pas l'air d'être au parterre pour voir figurer les maîtres: qu'on n'y voie plus de ces décorations de théâtre, de ces statues de plâtre.'[38] And Lequinio's observations on national fêtes include the wish that traditional divisions between actor and audience should be ignored there: 'Il faut que chacun se trouve acteur et spectateur tout en même temps.'[39]

On 18 floréal an II (7 May 1794) David presented the Convention nationale with his plan for the fête de l'Être Suprême. He envisaged the erection of a huge mountain bearing the tree of liberty on its summit, and on whose slopes the citizens would assemble to sing patriotic songs. David's instructions, which the Convention accepted, emphasize the involvement of ordinary men, women, and children in the ceremony. As their voices combine in the singing of a hymn to the fatherland, the mountain itself seems to pulse with life. Domestic and civic virtue commingle, with mothers hugging their suckling babes and holding up infant sons to the heavens in homage to the author of nature. Daughters, equally, cast into the skies their humbler offering of flowers, 'seule propriété dans un âge aussi tendre'. Meanwhile, sons hasten to prove their valour, drawing their swords, presenting them to their fathers, and swearing to use these weapons in the cause of upholding equality and liberty against the oppression of tyrants. Finally, all assembled seal their devotion in a fraternal embrace: 'ils n'ont plus qu'une voix, dont le cri général, *vive la République*, monte vers la divinité.'[40]

De Moy's work *Des fêtes* observes, in contrast, that while a fête is in general an action, a sort of drama in which the celebrants ('fêtants') are actors and the thing celebrated the dramatic subject, the public itself is the audience in the stalls. But there is a distinction to be drawn between such a display and stage drama. In fêtes, the subject always interests the actor himself, so that his role is natural and congenial to him, for

c'est sa propre joie ou sa propre tristesse qu'il manifeste; c'est son propre plaisir ou sa propre douleur qu'il nous peint; c'est de sa bonne ou mauvaise

---

[38] See Mona Ozouf, 'Le Simulacre et la fête révolutionnaire', *Les Fêtes de la Révolution: Colloque de Clermont-Ferrand, juin 1974* (Paris, 1977), p. 324.

[39] Joseph-Marie Lequinio, *Des fêtes nationales* (Paris, n.d.), p. 12.

[40] *Procès-verbaux du Comité d'instruction publique de la Convention nationale*, ed. J. Guillaume, 7 vols. (Paris, 1891–1907), IV. 347–50.

fortune qu'il nous entretient; enfin, c'est son bonheur qu'il publie ou son malheureux sort qu'il déplore et sur lequel il cherche à nous intéresser.[41]

The stage actor, on the contrary, is only momentarily involved in the character he embodies. What really interests him is the applause he wins from his audience and the fees they pay him: 'Voilà ce qui fait l'objet de son culte dans la fête qu'il nous donne; mais l'objet de sa fête à lui, c'est son salaire, c'est surtout le suffrage et les applaudissements des spectateurs' (p. 10). De Moy goes on to contrast 'public' and general fêtes. The latter he describes as games in which the public highway is the stage (p. 12), and where there are no idle spectators, as there are at public fêtes. The action is common to all, and every individual performs and participates. Lest his argument should still appear unclear, de Moy reiterates the point: 'Dans une fête générale, tous les individus sont donc acteurs, ou censés l'être; les rôles sont partagés d'avance; chacun, ou du moins chaque classe d'individus, doit y jouer le sien' (ibid.).[42] Finally, he writes, when the general fête is celebrated by the whole social body, the entire republic is not too large for the number of actors involved and the area which the action should cover (p. 13).

A further pointer to the prestige of acting during this time may be discerned in the fact that actors themselves were often revered as the mouthpieces of Revolutionary sentiments. Sometimes these sentiments were contained in plays written well before the Revolution, such as Corneille's *Cinna* and Voltaire's *Brutus*, but the existence of such works was commonly taken to indicate that the republican feelings they proclaimed were native to the French character, and therefore deserved to be expressed on the national stage. The government arranged free performances of such dramas for those citizens who could not afford the price of a theatre ticket.[43] The public adulation of the actor Talma, particularly after he left the Comédie-Française (which in the course of the Revolution became the Théâtre de la

[41] [Charles-Alexandre de Moy,] *Des fêtes, ou quelques idées d'un citoyen français relativement aux fêtes publiques et à un culte national* (Paris, an VII), pp. 9–10.

[42] Noverre, on the other hand, wonders whether a people can be at once actor and spectator at a fête (*Lettres sur les arts imitateurs en général, et sur la danse en particulier*, 2 vols. (Paris, 1807), II. 251).

[43] The Comité de salut public decreed that from 4 August to 1 September 1793 the tragedies *Brutus*, *Guillaume Tell*, and *Caïus Gracchus*, together with others which retraced 'les glorieux événements de la Révolution, et les vertus des défenseurs de la liberté', should be staged three times a week in the theatres of Paris, one performance being paid for by the republic (*Procès-verbaux du Comité d'instruction publique de la Convention nationale*, II. 688).

Nation) for the more progressive Théâtre de la République, was certainly connected with his performance of 'revolutionary' parts in dramas.[44]

Another factor explaining the special favour extended to performed drama and actors over this period was the common assumption that the theatre worked its effect through a union of sense-impressions, and could thus be argued by adherents of sensationalist philosophy to arouse more thoughts and feelings in the audience than other arts which appealed to a single sense. But at the same time upholders of sensationalism, as well as others, could reason that drama which involved a high degree of *actio* and other visual elements was bound to make a particularly forceful impression on the spectator. Then there was a particular advantage to theatres and actors in a further circumstance which accompanied the Revolution, namely the abolition of the privileges enjoyed under the old order by the royal troupes and the consequent opportunity for freely staging all kinds of dramatic spectacle. A wider audience than had been possible before was thus available for dramatic performances in a larger number of theatres than had been able officially to exist prior to 1789.

But there is a danger, which I have tried to avoid, in presenting too schematic a picture of discussions about acting, and particularly *actio*, during the second half of the eighteenth century. A desire to demonstrate the academic respectability of the profession certainly lay behind the attempt to create training schools for actors, and to reduce the principles of acting to rule, which I examine in the final chapter. But some cultural developments of the age connected with the art of gesture and movement can scarcely be related to this question. Broadly speaking, the popularity of pantomime is one such development, and the revolutionary change in dancing associated with Noverre another. Noverre's creation of a balletic style in which a new emphasis was placed on depicting action did not arise from the desire

---

[44] The predominantly conservative troupe of the Comédie-Française was originally fearful of performing Marie-Joseph Chénier's 'revolutionary' play *Charles IX*, but was eventually persuaded to do so. Talma played the main part, and became identified with the sentiments it conveyed by the public; and his relations with other members of the company were consequently strained. Some of them alleged that he had been carried through the streets to the theatre by the people, 'au milieu des cris de "vive Talma, le bon patriote; les comédiens sont des aristocrates qu'il faut mettre à la lanterne!"' According to Talma, the Comédiens' statement was false. See the *Exposé de la conduite et des torts du Sieur Talma envers les Comédiens Français* (Paris, 1790). A copy held by the Bibliothèque nationale has manuscript annotations by Talma, one of which conveys the above-mentioned denial.

to confer the dignity of classical precedent on dance by invoking the rhetorical art of *actio*. Rather, it resulted simply from an impatience with the nature of physical movement in ballet as it was then generally known. Noverre deplored mere virtuosity in dancing for reasons similar to those which lay behind contemporary disaffection from the 'academic' aspects of painting and sculpture. Diderot, among many others, berated the deadening preoccupation with technique which he found in some contemporary painters, and argued against his sculptor friend Falconet that idea ought to be granted an importance equal to, or greater than, that accorded to execution in art.[45] In like fashion Noverre criticized the cultivation of purely technical accomplishment that contributed nothing to a larger artistic conception, such as the narrative unfolding of his action ballets. This latter, in his view, constituted the proper object of the dancer's skills. *Actio*, as the name of the new balletic genre implies, was an essential element in their creation, as it is the very fabric of dance itself, but Noverre did not emphasize it in the spirit of the ancient rhetoricians. He believed that the presentation of *isolated* action impoverished dance. As Diderot encouraged painters to free themselves from the 'limiting obsession with posed academic models, and instead to attend to the movements, and configurations of life in the street, so Noverre exhorted dancers to be guided by the living, continuous action of the everyday world, and to let the observation of natural rather than contrived attitudes mould their performance on stage.

The conclusions indicated by the aspects of eighteenth-century dramatic performance discussed in this book are uncertain. By the end of the century the actor's lot in France certainly seemed better than at any previous time. In the mid-1790s, after the dissolution in 1793 of the royal academies created in the seventeenth century, an all-embracing academy for the arts and sciences, the Institut de France, was formed. The fine arts section of the Institut admitted actors, an elevation which finally compensated for what had seemed to many commentators a deplorable official neglect, not to say persecution, of the profession. But balanced against the new dignity was the fact that it proved to be temporary. When the actor Préville left Paris, and therefore had to become a non-resident associate, Grandmesnil was chosen to replace him; but he was the last actor to be

[45] See Diderot and Falconet, *Le Pour et le contre*, ed. Yves Benot (Paris, 1958), p. 187 ff., especially p. 202.

elected, and from 1803 his profession was excluded from membership. An edict of 23 January that year substituted four *classes* for the original three: there were sections for physical and mathematical science, the French language and literature, ancient and modern history, and the fine arts, but the art of declamation was no longer allowed its representatives (and this despite the fact that Napoleon was a friend and admirer of Talma's). Other evidence suggests, besides, that at the end of the eighteenth century imputations of 'actorliness' remained, as they remain today, often pejorative. Although the actor's 'infamie' had been lifted by law, and his friendship was cultivated by some of the highest in the land, the prejudice that had dogged him for centuries continued to weigh on the histrion's profession. Theory and then legal decree had established his right to social regard, but social practice in post-Revolutionary France still branded him an inferior citizen. For all the vigour with which actors and their supporters had argued the dignity and antiquity of their art, the stage performer was still associated in many minds with loose living, and his profession frequently regarded as lower in status than that of the scientist, writer, painter, musician, and even dancer.

CHAPTER ONE

Persuasion and the Visual Image

It is unsurprising that the age which developed theories about bodily eloquence in acting should have been one which assigned particular importance to the visual perception of reality. There were various reasons for this preoccupation. Whether they regarded the new emphasis on the actor's gesture and movement primarily as a product of restrictions that had been imposed on his activity by jealous official troupes or, more simply, as a legacy of the *Commedia* artistes to their followers in fairground and boulevard theatres, commentators on drama recurrently stressed one idea. What is seen, they asserted, affects more forcefully than what is perceived through the other senses; and in making this assertion they echoed many other writers of the day. Of course, it was open to critics to deny the proposition that visual images exercise such an appeal, and some did so. Yet the widespread concern with *actio* as it figured in performing arts of the period is some indication of the seriousness with which empiricist beliefs about sensory perception generally, and especially that of sight, were entertained in the age of Diderot. It would be rash, certainly, to discount the important of popular culture too in fuelling an interest in bodily eloquence and its effect on the onlooker in 'arts de spectacle', and I shall consider this matter in later chapters. Here I wish to concentrate on the philosopher's view of perception and the way it affected eighteenth-century theories about persuasion in the arts.

The notion that understanding is a function of sensation was an article of belief whose dissemination in the Enlightenment can be partly attributed to the empiricist philosophy of Locke. His influence may be felt in such works as the abbé Dubos's *Réflexions critiques sur la poésie et sur la peinture*, which in turn shaped much writing on aesthetics in France. Locke's assertion that the mind acquires its ideas through the senses lay behind Dubos's investigation in the *Réflexions* into the way in which the arts of poetry and painting exercise their persuasive power. According to Dubos, the arts induce their various forms of belief through the impression they make on the senses; and since all men are sensorily equipped, all are susceptible to the effects of art. Although Dubos refines this belief in the *Réflexions* to account for the possibility that an educated élite of connoisseurs may appreciate the arts more keenly than does mankind in general, his emphasis on the sensory rather than intellectual appeal of art allowed for a new view of artistic response which took account of common, 'popular' taste. Later writings built on the ideas to which the *Réflexions* gave currency to establish a conviction that the 'sensible' quality of art, or its capacity to arouse the feelings, was to be prized above its appeal to the rational faculty. Emphasis was then laid on the special power of visual perception as against that of the other senses.

The ability of the visual arts in particular to move the emotions had, of course, been stressed for centuries. Notably, it had been turned to account for the purpose of inducing belief in the principles of religion. The Jesuits (whose celebrated college dramatic productions made use of pantomime and other kinds of *actio*), and Catholics generally, emphasized the power of images as aids to devotion. They conceived of understanding as residing in the possession of correct images, and consequently saw the projection of the latter as the most direct means of achieving knowledge. In the *Spiritual Exercises* of Ignatius, the founder of the order, one exercise—the 'composition of place'—involved the participant's forming of mental images such as the Crib or the Cross as a form of contemplative piety. By this means, Ignatius argued, the worshipper would feel their devotional significance. Protestants often deplored this recourse to the seen on the grounds that representing the divine otherwise than through words was close to idolatry. The Trinity, for example, was frequently spoken of as literally 'unimaginable'. Even the Catholic Bossuet, writing about the 'catéchisme des fêtes', noted that

au lieu que dans les autres fêtes dont le mystère s'est accompli visiblement on peut concilier l'attention par des images qu'on en donne, quand il s'agit de parler de la Divinité, ou d'appliquer la Trinité adorable, on doit commencer à rendre le peuple attentif en lui faisant remarquer qu'en cette fête on ne lui propose aucune image sensible, parce que ce qui regarde la Divinité et la Trinité des personnes est tout à fait au-dessus des sens et de l'intelligence humaine.[1]

The potency of visual art had also been used in France as in other European countries to inspire belief in and reverence for the institution

---

[1] Quoted in Antoine Arnauld's *Réflexions sur l'éloquence des prédicateurs* (Amsterdam, 1695), pp. 83-4.

of monarchy. The creation in seventeenth-century France of an official academy for painting and sculpture led to an encouragement of works celebrating royalty which was later criticized, especially during the Revolutionary period, as a deplorable product of institutional conservatism. Both these examples—the use of images in religious worship and their deployment to social and political ends—indicate that empiricist philosophers were far from original in arguing the power exercised by what is seen. None the less, the broad transmission of this idea over the eighteenth century was heavily indebted to current philosophical notions.

Then, the relation between human passions and the perceptions of the different senses was invoked in connection with the power of images in art. There was nothing new, however, in the suggestion that the arousal of emotion was an essential part of aesthetic experience; for it had featured prominently in the poetic theory and philosophy of the ancients. Plato had laid against dramatic poetry the charge that it corrupted reason (a charge that Aristotle was subsequently to contest), arguing that emotion itself was an affliction, opposed to reason and inimical to thoughtful judgement.[2] In the *Republic* Plato accused the authors of tragedy and comedy of playing on feelings which were not open to reason, and their productions of arousing the emotions in a non-cognitive fashion. As we have seen, visual appeal such as later writers came to associate with a performer's bodily action in the theatre might be thought particularly dangerous in virtue of the facility with which the seen affects men's emotions. But for Aristotle intellectual cognition was itself involved in emotional response, and this meant that the latter could be distinguished from simple bodily sensations and drives. As a consequence of Aristotle's assertion, emotional appeal acquired dignity within the theory of rhetoric, and could thus be emphasized as a legitimate aspect of the orator's activity. With the later application of rhetorical theory to other arts, their effect too was positively discussed in terms of passionate arousal.

When eighteenth-century philosophers examined the passions, their conclusions were often influenced by the Aristotelian belief that their intellectual component, even their intellectual origin, gave these passions dignity. But they also investigated the arousing of emotion by an uninvolved agent. The 'cold' actor in Diderot's *Paradoxe sur le comédien*, for example, is able to move his audience by displaying

[2] See W. W. Fortenbaugh, *Aristotle on Emotion* (London, 1975), p. 18.

feelings which are governed by reason, and whose effect can be calculated and controlled. In this dialogue Diderot advances the view that the actor who lacks sensibility is greater than the one who involves himself emotionally in his role: the former's activity is guided by reflection and observation and can be sustained at will, whereas the latter's depends on a state of empathy which may vary from one performance to the next. It is the controlled performer and not the actor ruled by passion who wins his audience over and makes it his creature.[3] Sensibility, according to the *Paradoxe*, is not a characteristic of true genius. The actor 's'écoute au moment où il vous trouble, et . . . tout son talent consiste non pas à sentir, comme vous le supposez, mais à rendre si scrupuleusement les signes extérieurs du sentiment que vous vous y trompiez' (p. 312).[4] (Whether the audience's own emotions are similarly governed by the rational faculty is a question Diderot does not raise in the *Paradoxe*.) In seventeenth- and eighteenth-century rhetorics which discuss the actor's art as well as the preacher's and the advocate's, emphasis is often laid on the need to check both the performer's and the audience's emotions for the most effective use of persuasion to be made. Often, as we shall see in the next chapter, the desirability of such limitation was thought to entail the strict control of the speaker's *actio*, although such control was commonly regarded as more necessary in the pulpit and law-court than on the stage.

Emotional responses, for Aristotle, were neither bodily sensations nor bodily drives, but part of a group of sensations belonging to the soul. The catharsis of emotions which Aristotle believed tragedy to effect was not merely physiological, but had moral repercussions too.[5] In his *Réflexions* Dubos implied that dramatic art was instructive, rather than corruptive of reason, in making audiences aware of the nature of their emotions, and that it did this by creating a dramatic distance between the spectator and the characters depicted. Thus, according to Dubos, Racine's imitation of Phèdre's death allows the

---

[3] *Paradoxe sur le comédien*, in *Œ*, p. 309. Subsequent references to the *Paradoxe* are to this edition.

[4] This view contrasts with that expressed in the *Encyclopédie* article 'Génie', attributed to Saint-Lambert but in whose writing Diderot is generally supposed to have had a part. According to this article, the man of genius is governed by passions stronger than those of other mortals, and is dominated by 'enthousiasme'.

[5] See Jacob Bernays, 'Aristotle on the Effect of Tragedy', *Articles on Aristotle, 4: Psychology and Aesthetics*, ed. Jonathan Barnes, Malcolm Schofield, and Richard Sorabji (London, 1979), p. 156.

spectator to enjoy the emotion aroused in him without making him fear that it will be lasting (I. 28). The affliction is superficial, and will disappear with the fall of the curtain. Although the man experiencing emotion, such as the protagonist of a play, may not be in a position to understand its meaning, the detached observer is able to do so (e.g. I. 631). Thus drama may 'purify' the passions, in this technical sense of catharsis. When Dubos remarked in the *Réflexions* that the power of art to impress its audience is essentially dependent on that audience's detachment from the occasion of passion (I. 23, 25 ff.), he was arguing within the Aristotelian tradition. Dramatic art pleases us by allowing us to observe emotion without painfully experiencing it.

For 'sentimentalists' later in the eighteenth century, admittedly, such detachedness might well have seemed both impossible and undesirable. The same Diderot who praised the artist's self-control in the *Paradoxe* declared elsewhere that consumers of art should give themselves over unreservedly to the emotions aroused in them by plays, pictures, and novels, the more completely to respond in practical terms to the humanitarian impulses which art, according to Diderot, should provoke. But in eighteenth-century France sentimentalists could also be rationalists: Diderot himself exhibited such a dualism. It enabled him both to associate himself with the *Encyclopédie* article on 'génie' which extolled abandonment of the self to art and to write the fragment 'Sur le génie' which, like the *Paradoxe*, prized the artist's uninvolved and observant spirit above such abandonment. For Diderot the bourgeois dramatist the audience's emotional response to drama was desirable because the characters in whose lives it was meant to involve itself were predominantly sympathetic, so that the arousing of passion had beneficial rather than harmful consequences. Rousseau, however, argued a different view in his *Lettre à M. d'Alembert sur les spectacles* of 1758, in which he suggested that the impulse to good action that a play may stimulate is fleeting—no more than the homage that vice pays to virtue, and never converted into beneficent deeds in real life (pp. 33–4). Dubos also envisaged this possibility, but chose not to emphasize it.

The importance of stirring the emotions if the effect of art (its inducing of belief) was to be achieved had also been stressed by the orators of antiquity,[6] for whom the moral purpose of the rhetor's art took precedence over its artistic appeal. The latter simply facilitated

[6] See Cicero, *De oratore*, I.v.17, viii.31, xii.53, xv.67 (cf. Aristotle, *Rhetoric*, 2.i–xi), III.xiv.54; *Orator*, xxxvii.130–xxxviii.131; xl.138; *Brutus*, xlix.185.

the former, and remained subordinated to it. The orator appeared to experience passion as he tried to move the audience, but remained in some sense untouched by it, like the cold actor of Diderot's *Paradoxe*. Although, as one eighteenth-century commentator had it, rhetoric taught the arousing of emotions, it might govern their effect through its rational character.[7] Knowledge of the passions was required, in the classical orator and the imitators of his art in the pulpit, at the Bar, or on the dramatic stage, to ensure the moving of the audience. It was argued that this obtained even when the speaker's intent was to deflect his listeners from the path of countervailing passions,[8] or to purify them. Limitations were placed on the bodily expression of emotion on the part of the orator or actor, in case excessive physical portrayal might be taken to indicate that the line between depicting and undergoing that emotion had been crossed. Here an underlying assumption seems to have been that if he was actually possessed by passion, his consequent lack of rationality might debar him from being a sufficiently objective guide to its nature. As far as the *actio* of the orator was concerned, opinions about the physical expression of passions and its importance for transferring them from speaker to audience rested, in the eighteenth century as in antiquity, on the assumption that a direct relation existed between outward show and inward, passionate (or apparently passionate) state. Thus attention was focused on the bodily transmission of sentiment, or the physiology of persuasion.

For Dubos, it was axiomatic that man suffers more from living without passions than from experiencing them, and that he constantly seeks the means of gratifying his desire for emotion (I. 11). Art allows him to exercise his passionate impulses without directly and painfully involving him, for art is an illusion which counterfeits human experience harmlessly. Poetry and painting imitate events in the world of a kind which inspire the observer with emotion as he contemplates them, but the imitations have no unpleasant consequences. In life the experience of emotion results, characteristically, in action; in art the consequence is conviction. This property of art makes it desirable, not just permissible, that it should take as its subject things which engage the feelings rather than leaving men indifferent, and which for this reason must bear an evident relation to human life. According to

---

[7] J. P. Papon, *L'Art du poète et de l'orateur*, 6th edition (Paris, 1806), p. 154.
[8] Villiers, *L'Art de prêcher* (in Dinouart, *L'Éloquence du corps*, 2nd edition (Paris, 1761), pp. 352–3).

Dubos, the landscapes of Titian or Carracci, however great the artists themselves, fail to touch the passions because of their apparent isolation from the human. Great painters, mindful of the need to involve the spectator, ensure his attention by introducing human figures into their compositions: 'Le paysage que le Poussin a peint plusieurs fois, et qui s'appelle communément l'Arcadie, ne serait pas si vanté s'il était sans figures' (I. 50).

Dubos's emphasis in 1719 on the potential involvement of all men in works of art was repeated by subsequent eighteenth-century thinkers. During the Revolution the political leaders of the day gave practical expression to their belief that drama made an especially powerful appeal to the emotions by organizing free public performances of 'republican' plays like Marie-Joseph Chénier's *Caïus Gracchus* and Voltaire's *Brutus*. But earlier in the eighteenth century, and despite the egalitarian impulses central to the theories under discussion, difficulties began to appear concerning the doctrine of sensationalism. The notion of a hierarchy of knowledge which had previously been expressed in terms of a qualitative distinction between reason and sensation was no longer acceptable. The need to distinguish between qualities of belief therefore gave rise to theories about the different types of sensation experienced by men. In Dinouart's *L'Éloquence du corps, ou l'action du prédicateur* (1761), where the rhetorical concept of *pronuntiatio* is employed in the extended sense of bodily as well as verbal eloquence, the conventional opinion is expressed that as most of the preacher's audience will be of limited intelligence, they will require the stimulation of their senses for indoctrination to be effected. But although ordinary mortals are to be led 'plus par les sens que par l'esprit' (p. 15), the sense of sight is to be regarded as procuring the most immediate understanding, and the speaker's bodily eloquence is therefore a crucial contributory element in the impression he makes. In somewhat similar fashion to Dinouart, Marmontel observes in his article on the '[éloquence de la] chaire' that when the preacher in the pulpit addresses a great multitude, his words should themselves be 'sensibles' to be persuasive, and this is taken to mean that they must be full of images, tableaux, and movement.[9]

But before examining the notion that words may operate in a pictorial fashion, which Marmontel's comment implies, it is worth pursuing the larger assumption he makes concerning the directness of visual impressions as a factor in compelling belief. Dubos recalls

[9] Marmontel, *Éléments de littérature*, 3 vols. (Paris, 1879), I. 260.

Quintilian's reporting that the plaintiffs in court cases often had pictures of the crime for which they sought punishment exhibited before judges in order to incite the latter against the criminal (I. 35–6). In later writings emphasis was laid on the fact that pictorial images— at least where art is taken to be straightforwardly representational, as it was in eighteenth-century France—require no 'translation' to be understood. The directness of visual art was thus contrasted with the indirectness of verbal language, whose character is purely conventional. This notion too had its origin in ancient philosophy: it appears, for example, in Plato's *Cratylus*. Many moralists believed that such directness could have regrettable consequences. For Diderot and others, the dangerous facility of the painter to bring his subject alive before the beholder, and tempt him to take it for real, had constantly to be exposed for what it was. In 1707 the Dutch painter Lairesse had drawn attention to this problem in *Het Groot Schilderboek*, where he argued that the sight of obscene paintings is more pernicious than are equivalent obscenities in a different medium. This, he said, is because the eyes are the most efficient carriers of sensation, especially where erotic ideas are in question.[10] Diderot, whose writings on art are known to have been influenced by Lairesse, repeats the charge in his *Pensées détachées sur la peinture*, where he asserts that 'un tableau, une statue licencieuse est peut-être plus dangereuse qu'un mauvais livre; la première est plus voisine de la chose.'[11] Similar reservations are expressed by some commentators on *actio* in the pulpit, law court, and political assembly, and even in the theatre.

Dubos stated firmly that sight is the sense in whose testimony the soul has the greatest confidence (I. 375), and that pictures act more promptly on men's sensibility than other forms of art. The most effective way of inspiring belief in men or provoking their imagination, he declares, is to make them exercise their eyes or their mind's eye, so that pictures 'ne livre[nt] qu'un assaut à notre âme'. For Dom Sensaric, who in 1758 published a three-volume work on *L'Art de peindre à l'esprit*, visual means provided the orator with the most attractive method of achieving his persuasive ends, and were 'le plus sûr moyen d'éviter la sécheresse et par conséquent l'ennui'.[12] But his reference here is to the use of pictorial effects in or through language,

---

[10] Gérard de Lairesse, *Le Grand Livre des peintres*, 2 vols. (Paris, 1787), I. 200.
[11] *Pensée détachées sur la peinture*, in Œ, p. 769. Subsequent references to the *Pensées détachées* are to this edition.
[12] [Dom Sensaric,] *L'Art de peindre à l'esprit*, 3 vols. (Paris, 1758), I. i.

rather than to the speaker's supplementing his speech by recourse to *actio*.

There had been a tradition linking words and images since classical antiquity. Lairesse, among many others, mentions that the use of rhetorical figures in verbal discourse, like the graphic quality of a painting, can impress men with the physical reality of what is described, for the weak human intelligence cannot grasp purely intellectual notions. Men take percepts for reality: to be, for them, is to be perceived. Dubos refers to Quintilian's ability, in the course of a speech, to conjure up the subject of his discourse for his listeners, so that they actually saw it before them. In the seventeenth century Bernard Lamy, following in this tradition, defined the verbal figure of the trope as a 'peinture sensible de la chose dont on parle', and recalled that the name of *hypotyposis* was given to the technique by which a speaker brings an object so clearly to his audience's mind that the latter seems to see what is being said.[13] The classical concept of *enargeia*, or vividness, was applied by rhetoricians to verbal art, as Aristotle had used it to describe the graphic quality of a successful play when it is merely read rather than being performed.

Much of the debate about the propriety of rhetoric in the seventeenth and eighteenth centuries centred on the inclusion of 'images sensibles' in oratory, especially in preaching. Gibert, in his debate with François Lamy about the *ornatus*, declared against his adversary that such images were useful in impressing the reality of a speaker's subject on his listeners, and contended that nothing was so effective in combatting evil habits as to present them ('mettre en scène') with the help of metaphors, comparison, hyperbole, and other such figures.[14] While believing that images should be used sparingly, especially in didactic works (where they might confer a 'caractère de vanité', ibid., p. 18), Gibert argued that they may lend delicacy, power, nobility, beauty, and lucidity as well as vividness, when sensitively used (ibid., p. 12).

This theory of the respective relationship with the world of images and words had important consequences for the theory and practice of the arts. Both the Cartesian and the empiricist traditions of philosophy emphasized the role of ideas in understanding, and language

[13] [Père Bernard] Lamy, *De l'art de parler*, 2nd edition (Paris, 1676), p. 74, pp. 89–90; see also Pierre Zoberman, 'Voir, savoir, parler: la rhétorique et la vision', *XVIIe Siècle*, 133 (1981), 410 ff.

[14] Balthasar Gibert, *Réflexions sur la rhétorique, où l'on répond aux objections du Père Lamy, bénédictin*, 3 vols. (Paris, 1705–7), III. 10.

itself was thought to work in an 'ideational' way. Clarity and distinctness of image became the criteria of certainty and therefore of knowledge. For Locke and Hobbes, as for Descartes, the role of a word was the evocation of an image; and this notion was developed in the theory of rhetoric and of artistic value generally. In the case of drama, it is probable that such philosophical beliefs (at least as held by a philosopher–playwright like Diderot) were turned to account in theatrical theory and practice, and particularly in the stage presentation of abstract notions such as those relative to morality. The language was taken to work eidetically, and the natural consequence of this was for the importance of direct, visual imagery to be emphasized. Such a hypothesis goes some way to explaining what has usually been accounted a paradox in Diderot's dramatic work, that on the one hand his *drames* are periphrastic and prolix, while on the other the theory on which they are based stresses the need for less verbosity and more pantomimic action on stage.

Dom Sensaric contended that poets who were not painters were merely versifiers (I. iii), and by the same token orators who lacked the painter's talents were 'discoureurs', or at the most cold and abstract 'raisonneurs'. This last observation is a telling commentary on the disjunction between reason and sensibility perceived by many eighteenth-century thinkers, and again recalls the long-established belief that words as conventionally used appeal to the intellect above all else, whereas the senses refer men to the material world and empirical reality. According to Sensaric, while the adornment of speech with pictorial images must be strictly controlled and the temptation to prefer 'le brillant' to 'le solide' resisted, yet it should be recognized that the greatest orators of ancient and modern times took advantage of the clarity that visual elements could lend to their discourse. In the speeches of Demosthenes, Cicero, Bossuet, and Boileau, 'tout est image; tout est peint d'après nature; ils parlent autant à l'imagination qu'à la raison; ils élèvent l'âme et plaisent à l'esprit tout à la fois' (I. iv). Significantly, the author here records that all these men avoided the timid, over-refined art of the miniaturist. The implication is that they painted in free, broad brush-strokes, and is reminiscent of the comparison Horace draws in the *Ars poetica* between types of oratory and types of painting. Writing of the equivalence between *pictura* and *poesis*, he contrasts the speaker in a courtroom (whose speech demands close attention), or the painter whose pictures require close scrutiny, with the orator in a public forum (whose speech is heard at a distance and lacks minute finish), or the painter whose pictures are executed with a bold, rough hand, and should be contemplated from afar.

Many contemporaries agreed that, just as language could be more or less evocative according to the speaker's or writer's use of it, so forms of visual depiction were more or less emphatically direct in their statements. The latitude for interpretation offered by a sketch was often stated to be far greater than that available from a finished painting. This perception belonged to a tradition long established in the visual arts, but which acquired special power in the eighteenth century because of the taste for spontaneity and the *non finito*, which allowed wide scope for the play of imagination. In this respect a parallel might be drawn with the essential indirectness of verbal language. Some eighteenth-century thinkers indeed developed this notion into a theory of the crucial difference between most kinds of 'finished' visual art and most arts of the word, where the emphasis was on the ability of words to suggest rather than state outright. As we shall see, a fondness for exercising the imagination also lay behind the preference of many critics of the time for a non-verbal, or only partially verbal, art in theatrical performance, and hence led them to declare a taste for pantomime.

The belief that literature operates in a manner fundamentally different from that of the visual arts (where the essential clarity, rather than ambiguity, of the visual image is at issue) informs a number of writings on aesthetics during this period. Some of the arguments relating to this topic concern the non-temporal depictive possibilities of painting and sculpture as against the temporal description and analysis available to the writer. Of particular importance here is the belief, developed by Burke and then by Lessing, that literature does not seek the directness of the visual arts.[15] The chief preoccupation of literature, rather, is with inward qualities: moods, thoughts, and other intangible states. Burke argues in the *Enquiry into the Origin of our Ideas of the Sublime and Beautiful* that literature does not communicate through images, as visual art does, but suggests the effect of things without a clear picture having been presented.[16] (Painting may do this, as I shall suggest later, through devices such as symbolism.) Burke therefore denies that poetry is an imitative art, in the sense in

---

[15] See E. Allen McCormick, '*Poema pictura loquens*: Literary Pictorialism and the Psychology of Landscape', *Comparative Literature Studies*, 13 (1976), 199.

[16] Gita May, 'Diderot and Burke: A Study in Aesthetic Affinity', *PMLA*, 75 (1960), 536.

which painting and sculpture are imitative, and suggests that it affects through sympathy. Verbal description can, however, convey affections; and when this is well done, words work their effect without presenting the reader or listener with any precise idea of the object or objects which gave rise to the emotion. For Burke, an attempt to render reality in detail with words would hinder the effort at conviction by drawing attention to detail. (This argument again recalls that of Horace's *Ars poetica*, discussed above.) In this connection Burke quotes the description of Helen in the third book of the *Iliad* in order to show that good poetry conveys its subject by rhythm and sound, and always leaves something unsaid (May, art. cit.).

Much was made by contemporaries of the belief that painting affects the onlooker instantaneously, whereas literature operates through time and cannot be comprehended in a single moment. Some of the ways in which eighteenth-century French painters nevertheless resisted the temporal limitations of their medium will be noted in a later chapter. But some writers and artists denied that painting could be understood instantaneously; and where this view was advanced, it was sometimes done for the purpose of elevating painting into an art not merely of surface depiction, but bearing a significance which had to be gradually grasped by patient investigation of its hidden meaning. Obviously, painting that employed allegory or symbolism might thus frustrate a superficial effort at understanding, but a weakness of such art, as its critics declared, was that its reference might resist every attempt at elucidation. Painting which guided the observer progressively towards full understanding, on the other hand, was highly prized, because its meaning was many-layered rather than arcane. The anonymous author of a letter to the *Journal de Paris* in 1791 observes that David's painting of the *Serment du Jeu de Paume* exercised its particular powerful effect by refusing to allow the spectator immediate comprehension of its subject.[17] Its intention was to show the history of the Revolution, although its subject appeared confined to the first period thereof, and was ostensibly a representation of a single event at Versailles (*see Plate 2*). According to the writer, David had had recourse to an art resembling that practised by the great poets of antiquity, and had linked to his subject, by well-chosen episodes, the most honourable aspects of eighteenth-century philosophy, signalling the progress of reason and enlightenment. As we shall see, supporters

---

[17] MS *Lettre aux auteurs du 'Journal de Paris'*, in Bibliothèque nationale, *Collection Deloynes* (subsequently referred to as *Deloynes*), vol. XVII, no. 435.

of pantomime in France, who often compared that art with painting, sometimes mentioned the former's ability similarly to suggest rather than state outright, and valued it correspondingly. On stage as on the canvas, the seen could provoke the imagination rather than constrict it, and lead the spectator's thoughts onward to comprehension.

Diderot's ideas about the visual apprehension of reality are intriguingly described in the *Éléments de physiologie*. His view of the manner in which phenomena are perceived is described in detail, and the main burden of the argument is, again, that visual images are grasped successively through time rather than instantaneously. In order to have an exact notion of the parts and the whole of a tree, for example, one must first turn one's attention inside oneself: the imagination depicts all on the 'inner eye', and 'l'on procède au-dedans de soi . . . par champs plus ou moins étendus.'[18] This, Diderot emphasizes, is not a static process of image-taking, but a sequential one, for Diderot's interest is in the physiology rather than the anatomy of perception. Such a manner of seeing, in which the subject's imagination combines with his perception of objects over time, brings to mind Diderot's description of watching actors in the *Lettre sur les sourds et muets*, where he calls *actio* more evocative than words. It also recalls to the reader the fact that Diderot, like Noverre, preferred the live model in visual or performing art to the academically posed and lifeless one, and suggests that his theory of perception and his aesthetic sensibility had a common basis.

The process of perception, according to the *Éléments de physiologie*, may be effected so rapidly that the observer is convinced he sees the whole tree at once inside himself, as he believed he did outside himself, which, according to Diderot, is true in neither case. The same holds for the perception of abstract entities such as truth—a main preoccupation of Diderot's *drames*, and, as I shall suggest, a major reason for the concentration on visual impressions evident in the *Entretiens sur 'Le Fils naturel'*. Such abstract qualities can only be known through the subject's proceeding systematically, and even then 'La vérité peut tenir tellement à l'image totale qu'on ne puisse ni affirmer ni nier d'après le détail le plus rigoureux des parties' (p. 235). But again: 'Il n'y a qu'un moyen de connaître la vérité, c'est de ne procéder que par partie, et de ne conclure qu'après une énumération exacte et entière' (ibid.). The progress of the mind,

---

[18] Diderot, *Éléments de physiologie*, ed. Jean Meyer (Paris, 1964), p. 226.

1 Ziesenis, Le 'Père de famille' de Diderot. Photo: Bibliothèque nationale

2 After David, *Le Serment du Jeu de Paume*. Photo: Bulloz

3 Moreau le Jeune, *Mirabeau arrive aux Champs Élysées*. Photo: Bibliothèque nationale

4 David, *Les Licteurs apportant à Brutus les corps de ses fils*. Photo: Bulloz

Diderot also notes at this point, is merely a succession of experiences (although 'La promptitude est la caractéristique du génie', p. 236). Again, the reader is put in mind of the successive revelations effected by the bodily eloquence of the actor. The visual image, like that presented in narrative paintings by Diderot's acquaintance Greuze, tells a story whose burden cannot be grasped immediately.

Significantly, all Diderot's references to imagination in this work are expressed in terms of sight, as though he himself were convinced of its primary importance in mental life. The definition he offers at one point is a wholly visual one: imagination is the 'faculté de se peindre les objets absents comme s'ils étaient présents, d'emprunter des objets sensibles des images qui servent de comparaison, d'attacher à un mot abstrait un corps' (p. 250). Whereas imagination is associated with objects, memory is merely of signs (whose mode of reference, in Diderot's view, may be purely conventional). Further on he calls imagination 'l'œil intérieur', and observes that 'la mesure des imaginations est relative à la mesure de la vue' (ibid.). The blind, according to Diderot, may 'image' even though they cannot see: '[ils] ont de l'imagination, parce que le vice n'est que dans la rétine.' Finally, Diderot draws an analogy with graphic art as he writes that imagination could be measured by comparing two drawings, equally skilful, of the same object. Each of the artists 'se fera un modèle différent, selon son œil intérieur et son imagination, et son œil extérieur. Les dessins seront entre eux comme les deux organes' (p. 251). From all this, and from much supporting evidence in his other writings, it is clear that Diderot's own imagination is a strongly pictorial one. Indeed, he observes in the *Salons* that the artists he knows have congratulated him on his perceptiveness in this respect: 'Chardin, Lagrenée, Greuze et d'autres m'ont assuré, et les artistes ne flattent point les littérateurs, que j'étais presque le seul d'entre ceux-ci dont les images pouvaient passer sur la toile presque comme elles étaient ordonnées dans ma tête' (*Salons*, III. 109). The same *Salon*, that of 1767, also puts the argument differently, with Diderot claiming that his literary imagination has been shaped by the visual arts rather than the other way about:

mon imagination s'est assujettie de longue main aux véritables règles de l'art, à force d'en regarder les productions; . . . j'ai pris l'habitude d'arranger mes figures dans ma tête comme si elles étaient sur la toile; . . . peut-être [que] je les y transporte, et que c'est sur un grand mur que je regarde, quand j'écris. (Ibid., p. 110)

As I shall suggest in chapter 3, the imagination of another theorist of drama and occasional commentator on the visual arts, L.-S. Mercier, was also a pictorial one, at least to judge by the number of analogies drawn between painting and drama in his treatise *Du théâtre* (1773). It is curious, then, to note the antagonism towards the former art which he declared in a series of contributions to the *Journal de Paris* in 1797. Mercier's new view is that literature is superior to painting because of the latter's assertive character, which makes its appeal to the imagination less than that of words. 'La peinture est un enfantillage de l'esprit humain. La peinture n'existe que dans la langue écrite. Ces paroles resteront; que dis-je? elles vont fructifier. La génération qui s'éveille les entendra' (14 pluviôse an V/2 February 1797, p. 537). The argument is a familiar one: painting constricts the imagination, whereas the use of the intellect involved in reading and listening to words creates a wealth of images for man to consider and combine. Talking of a visit to a museum, Mercier observes that his own head is richer than 'cet immense cabinet'. All men can see what Mercier sees inside himself if they cultivate 'la peinture intérieure', whereas 'cette imitation grossière des choses créées n'est faite que pour ceux qui n'ont pas le sens intime, où la peinture intérieure déploie sa magnificence et ses richesses inépuisables et renaissantes' (ibid.). In subsequent numbers of the *Journal de Paris* Mercier continues his attack on the visual arts, and declares his belief in the greater analytical power of words. Initially he disdains to argue his case, simply stating that 'Le verbe, la parole, le discours' reproduce nature as it is, and show all other means up as false: 'le sauvage, quand il fait une harangue, est plus peintre que Raphaël' (10 fructidor an V/27 August 1797, p. 1399). But subsequently he reasons that painting is an imperfect medium for the conveying of reality because it fails to capture the successive quality of experience—a remark which is in the tradition of Shaftesbury's, Dubos's, and Lessing's aesthetic theory. (The ability of drama to do precisely this, on the other hand, was often noted by eighteenth-century commentators.) The desirability of the individual's using his own imaginative means to form a notion of reality is stressed by Mercier: 'pour exprimer parfaitement les objets de la nature, il faut des idées intellectuelles, qui n'isolent point les objets, parce que la reproduction de ces mêmes objets, et leur imitation parfaite, ne sont qu'en nous' (pp. 1399–1400). The clarity of painting, in other words, is intellectually deceptive and artificial. This opinion, as we have seen, can be related to the Platonic objection to art, which is surely at the origin of Mercier's own.

It is no surprise to find that Mercier values the sketch above the finished picture. He writes that the art of painting became impotent in proportion as it attempted to render its subject completely, whereas 'le dessin se rapproche, pour ainsi dire, de l'écriture; c'est moi qui fais le tableau.' Not altogether consistently, he avers that pantomime is to spoken drama as the sketch is to painting, and announces in the spirit of Diderot that the perfection of dramatic art is realized in pantomime. But this assertion drew a rejoinder from another contributor to the *Journal de Paris*, Villeterque, who remarked that 'L'acteur silencieux de la pantomime doit peindre les idées qu'il veut exprimer; mais s'il faut qu'il en ait, il ne sera supérier à l'auteur qui les écrit, ou à l'acteur qui les prononce, que lorsqu'il en aura davantage' (27 thermidor an VI/14 August 1798, p. 1372). It is difficult to believe, he continues, that actors in pantomime could have more ideas than Corneille, Racine, Voltaire, and Molière, whose plays Mercier does not wish to be performed. For Villeterque, pantomime is essentially limited (as it was for most writers from Diderot onward, who emphasized the need for it to be combined with words):

je crois qu'un acteur, si habile qu'il soit, ne laisserait pas que d'être un peu embarrassé s'il fallait, par exemple, qu'il produisît par son silence, l'expression de sa physionomie et ses gestes l'effet prodigieux qui, dans Shakespeare, résulte du beau discours d'Antoine après la mort de César. Cependant tout parle aux yeux dans cette situation; la robe ensanglantée de César, son corps étendu au bas de la tribune, l'effroi de ses amis, son testament qu'on montre au peuple, l'agitation des conjurés, la tumultueuse incertitude des spectateurs: on voit tout, et cependant on écoute.

Villeterque proceeds to attack Mercier's often-repeated notion that the sensations man creates himself are preferable to those he passively receives, and takes as an illustration a picture of Psyche by Gérard. (Mercier's own praise of Gérard's picture, confusingly, was also printed in the *Journal de Paris*.) Villeterque writes that the artist did well to carry his canvas beyond the stage of a sketch, for 'Si l'art de la peinture se bornait au dessin, pourrais-je imaginer tous les détails séduisants qui m'enchantent dans ce tableau?' He declares that 'Il faut être né peintre ou sculpteur pour se figurer des tableaux ou des statues; l'imagination seule pouvait, comme l'assure le citoyen Reicrem [i.e. Mercier], achever ce que l'artiste a ébauché, pourquoi s'arrêter en si beau chemin et ne pas dire que l'admirateur doit tout créer?' (ibid.) He concludes that only Mercier's theories prevent him from admitting his taste for great art: just as he admires Gérard's work, but feels obliged to criticize painting in general, so he is devoted to the productions of the great seventeenth-century French dramatists, yet is driven to deplore the performance of their plays. Villeterque does not point out the most glaring inconsistency in Mercier's argument for pantomime, that he had shortly before declared his preference for verbal art over visual.

The relevance of this discussion to the performance of plays lies in the fact that conventional drama (that is, drama with dialogue) consists of words and plastic images, and that the importance of the latter was increasingly stressed in the course of the eighteenth century. *Actio*, it was argued, gave added force to the words an actor declaimed. Sainte-Albine, writing about the art of the actor, remarked in 1747 that enacted plays present their audience with *all* the evidence, ocular and aural, that may effect persuasion or induce belief. He argued that men feel a complete pleasure in watching plays as conventionally performed because nothing needs to be supplemented by their imagination:

Quelque admirables que soient les ouvrages de cet art merveilleux [painting], ce ne sont que de simples apparences, et bientôt nous reconnaissons qu'il nous offre des fantômes pour des objets réels. En vain la peinture se vante de faire respirer la toile. Il ne sort de ses mains que des productions inanimées. La poésie dramatique fournit au contraire des idées et des sentiments aux êtres qu'elle enfante, et à l'aide du jeu théâtral elle leur prête la parole et l'action. Les yeux seuls sont séduits par la peinture. Les prestiges du théâtre subjuguent les yeux, les oreilles, l'esprit et le cœur . . . Son art est par cette raison un de ceux auxquels il appartient le plus de nous faire éprouver un plaisir complet. Notre imagination est presque toujours obligée de suppléer à l'impuissance des autres arts imitateurs de la nature. Celui du comédien n'exige de nous par lui-même aucun supplément.[19]

Whether this in fact makes drama more powerful as a rhetorical art than others whose resources are essentially limited to affecting a single sense seems doubtful, however. Quatremère de Quincy observed apropos of opera, against Dubos, that the arts which touch more than

[19] Rémond de Sainte-Albine, *Le Comédien* (Paris, 1747), pp. 14–15. Dubos also believed that the union of sense-impressions, as achieved in opera, increased the effect of a work on its audience (I. 65).

one sense may move the individual sense less powerfully than do those which stimulate only one. (I shall return to this point.) But for some eighteenth-century commentators the fact that drama gave living form to its characters meant that it had a special force in comparison with the visual arts to which it was evidently related. I have already mentioned Bossuet's opinion that performed drama is potentially more corrupting than painting whose subject is immoral, and Levesque's view that the same distinction exists between enacted poetry and poetry that is silently read. Neither writer considers the important possibility that man is most deeply moved by art in which he himself is actively involved, perhaps because they attach no particular value to the arousing of the imagination.

But, as we have seen, this quality of suggestion, of provoking thought, was precisely what Diderot admired in the mute performance of plays he describes in the *Lettre sur les sourds et muets*:

Le terme de jeu qui est propre au théâtre . . . me rappelle une expérience que j'ai faite quelquefois, et dont j'ai tiré plus de lumières sur les mouvements et les gestes que de toutes les lectures du monde. . . . Aussitôt que la toile était levée [au théâtre], et le moment venu où tous les autres spectateurs se disposaient à écouter, moi, je mettais mes doigts dans mes oreilles, non sans quelque étonnement de la part de ceux qui m'environnaient, et qui, ne me comprenant pas, me regardaient presque comme un insensé qui ne venait à la comédie que pour ne la pas entendre. Je m'embarrassais fort peu des jugements, et je me tenais opiniâtrement les oreilles bouchées, tant que l'action et le jeu de l'acteur me paraissaient d'accord avec le discours que je me rappelais. Je n'écoutais que quand j'étais dérouté par les gestes, ou que je croyais l'être. . . . Mais j'aime . . . vous parler de la nouvelle surprise où l'on ne manquait pas de tomber autour de moi, lorsqu'on me voyait répandre des larmes dans les endroits pathétiques, et toujours les oreilles bouchées. (pp. 52–3)

Two years later, in *The Analysis of Beauty*, Hogarth referred in similar terms to the force of action in dramatic performance when it was effectively divorced from words. He imagined a foreigner visiting an English theatre, completely ignorant of the language but conversant with all the resources of action and gesture. It is evident, he writes, that the man's sentiments under such conditions will chiefly arise from the movements of each character. Those of an old man, for example, will at once strike him as proper or inappropriate to his age, and he will similarly perceive low or odd characters according to the inelegant lines of movement which belong to Punch, Harlequin,

Pierrot, and the clown. Equally, the spectator will judge of the graceful acting of a gentleman or hero by the elegance of his movement in the 'serpentine line' which Hogarth proposes as a concrete key to pictorial beauty (p. 152).

Sainte-Albine's view of performed drama runs counter to Hogarth's, and to many influential theories of the age about the appeal of *actio* in the theatre. Even had it been based on fact, it would scarcely have recommended itself to a thinker like Diderot as a judgement of value. For the latter, as I have suggested, the greatest art was that which most powerfully provoked the imagination.[20] When the stage performer augments his verbal discourse with bodily eloquence, in his view, and so makes interpretative demands of the spectator, he increases the intensity of the latter's aesthetic experience; for Diderot believed with his age that the enjoyment of art depends on sensory as well as intellectual perceptions. Although this opinion was more firmly stated in the eighteenth century than at any previous time, it was not a product of that period. The philosophy of the ancients, and particularly Aristotle's answer to the Platonic view of art, had dealt with it in detail, but eighteenth-century thinkers drew on the conclusions of their forebears to develop a theory of artistic response that stressed the positive value of passionate arousal, which visual perception especially was believed to stimulate. They declared that the primacy of visual perception in this respect made artistic appeal a democratic affair, for what can be directly apprehended can be apprehended and judged by all. It was natural for Diderot, whose imagination was strongly visual, to emphasize in his own work the importance of what is seen (particularly in his drama, which he wanted to be performed before ordinary people, away from the bastion of privilege that the Comédie-Française represented), and attempt to introduce a new visual element into French acting. Others were equally convinced of the appeal exerted by performed drama containing images as well as words.

Where some contemporaries differed among themselves was in their opinion about the relative merits of drama and other art-forms. Sainte-Albine's view ran counter to Quatremère's, and would prob-

[20] In an addition to the *Lettre sur les sourds et muets* Diderot declared that music spoke more strongly to the soul than poetry or painting: 'Comment se fait-il . . . que des trois arts imitateurs de la nature, celui dont l'expression est la plus arbitraire et la moins précise parle le plus fortement à l'âme? Serait-ce que montrant moins les objets, il laisse plus de carrière à notre imagination, ou qu'ayant besoin de secousses pour être émus, la musique est plus propre que la peinture et la poésie à produire en nous cet effet tumultueux?' (p. 102)

ably have been contested by Diderot as well. While Saint-Albine prized the immediacy of conventional drama, which he took to be a function of its varied sensorial appeal, Diderot also set high value on art which worked by indirect means, and which invited, even necessitated, the beholder's active participation. This explains his fondness for pantomime, which suggests rather than states outright, and for the *non finito* generally. Furthermore, Diderot rejected the implication contained in Mercier's words that visual images are necessarily assertive and direct, and thus intellectually inferior to images conveyed by words. Rather, he attempted to show how all visual experience—in the theatre, the art gallery, or the outside world—is mediated, through different types of physiological equipment, and how perception through time renders the seen more complex than might at first appear. This opinion sheds light on Diderot's particular affection for the free-ranging images of *actio*. But the *Éléments de physiologie* shows that his interest in the mechanics of perception extended to other visual experiences than simply those provided by the theatre.

We know from the *Salons* as well as the *Essais sur la peinture* how Diderot deplored the static in art, and how he associated it with an unwillingness or inability on the artist's part to observe living models. For him this was the principal vice of the way 'life drawing' was practised at the Académie royale de peinture et de sculpture, where the student spent seven years observing and copying a posed and lifeless human figure. Instead, he wanted pupils to learn from a study of mobile forms, and thus to understand the sequential nature of visual perception. Some of the consequences of Diderot's preoccupation with movement in art, such as his extravagant enthusiasm for the teeming narrative canvases of Greuze, may strike the modern reader as regrettable. But in so far as they reflect his belief that in visual perception the mind and senses always pass through a succession of experiences, they are at least consistent with the theory that underlies his fascination with *actio*.

CHAPTER TWO

Action and Conviction

Many French rhetorics of the seventeenth and eighteenth centuries address their remarks to the actor as well as the preacher, barrister, magistrate, and member of the political assembly, but not necessarily approving the analogy between stage performance and the eloquence of the other parties. The aspect of rhetoric which deals with the speaker's bodily action is, according to many discussions of the subject, the one which the ancients most neglected to elucidate. Writers of the period regularly remark that Aristotle and Cicero had almost nothing to say about the orator's gesture and movement, but that Demosthenes had called *actio* the first, second, and third parts of eloquence.[1] Seventeenth- and eighteenth-century commentators repeatedly note that the only rhetorician of antiquity who properly investigated this branch of oratory was Quintilian.[2] Their own pronouncements on the subject, as a result, are often based on Quintilian's observations in the *Institutio oratoria*, but there are many attempts, especially in the second half of the eighteenth century, to move beyond classical doctrine. I have already touched on the developing concern in eighteenth-century France with the physical aspect of acting performance in the theatre, and shall return in later chapters to further consideration of why an emphasis on 'bodily expression' should have made itself felt at this time. Here I wish to examine the more general question of the relationship between acting and other kinds of performed eloquence (in the pulpit, the law-court, and the political assembly) from the point of view of bodily representation.

Luigi Riccoboni's *Pensées sur la déclamation* of 1738 defines the art of declamation as that of combining gestural expression with varied pronunciation, in order to increase the force of the underlying thought.[3]

---

[1] See, for example, Le Gras, *La Rhétorique française ou les préceptes de l'ancienne et vraie éloquence accommodés à l'usage des conversations et de la société civile: Du barreau et de la chaire* (Paris, 1671), p. 269.

[2] See, among others, Michel Le Faucheur, *Traité de l'action de l'orateur ou de la prononciation et du geste*, ed. and rev. by Conrart (Paris, 1657), p. 11. Le Faucheur observes that Quintilian's precepts were intended for practitioners at the Bar.

[3] Louis [Luigi] Riccoboni, *Pensées sur la déclamation* (Paris, 1738), p. 3.

He regards the masters of antiquity as models in this respect, saying that they have settled the main principles of the subject (p. 4), but that many moderns ignore their teaching. The perfect orator, he continues, combines verbal eloquence with 'external' eloquence; and the art of bodily declamation is an essential one to preachers, advocates, academicians, members of learned societies presenting the results of their research to an audience, conversationalists, writers reading their works to a public, actors, and others besides (pp. 6–7).

Writers of the time take it as axiomatic that the object of all eloquence, verbal or otherwise, is persuasion of the listener or beholder, and Cicero's observation that the orator aims to teach, please, and move[4] is echoed in many subsequent discussions of the way in which the rhetorical end is achieved. The extent to which each of these three aims should be pursued is seen as differing according to the type of eloquence in question. That the performance of drama is a type of eloquence is firmly stated by Luigi Riccoboni and by other eighteenth-century writers on the art of acting, but commentators vary in their willingness to see theatrical performance as comparable to declamation of a traditionally more respected kind. While Cicero's and Quintilian's rhetorics were principally intended for the training of advocates, French handbooks of the seventeenth and eighteenth centuries generally direct their advice at a wider field. Rapin, the seventeenth-century Jesuit, distinguishes two types of eloquence, that proper to affairs of state and that practised in the pulpit,[5] whereas Dubroca's treatise of 1802 speaks of three sorts, seen as being very different from one another: political, judicial, and ecclesiastical.[6]

The extension of traditional rhetorical doctrine to the art of acting, which is effected as early as 1708 in Grimarest's *Traité du récitatif*,[7] is recurrently tempered by the acknowledgement, explicit or not, that the latter is a late addition to the canon. Many writers describe actorliness in oratory as undesirable, however ready they are to grant *actio* a place in the total system of rhetoric. More often than not, evidence of 'theatricality' in a preacher's or advocate's delivery is taken to be a reason for condemning the style of his eloquence. I shall return to this

---

[4] Cicero, *Brutus* xlix. 185: 'Tria sunt enim, ut quidem ego sentio, quae sint efficienda dicendo: *ut doceatur is apud quem dicetur, ut delectetur, ut moveatur vehementius.*'

[5] [Père René Rapin,] *Réflexions sur l'usage de l'éloquence de ce temps* (Paris, 1671), p. 51.

[6] Louis Dubroca, *Principes raisonnés sur l'art de lire à haute voix, suivis de leur application particulière à la lecture des ouvrages d'éloquence et de poésie* (Paris, 1802), p. 403.

[7] See Peter France and Margaret McGowan, 'Autour du *Traité du récitatif* de Grimarest', *XVIIe Siècle*, 132 (1981), 303–17, especially p. 304.

important point in the next chapter with reference to the criticism of painting in eighteenth-century France, where theatricality of manner and subject is also commonly disapproved.

The anonymous author of an early eighteenth-century treatise on 'la bonne et solide prédication' (1701) adversely contrasts a theatrical type of preaching with what he regards as the truer, simpler method proper to the pulpit. Interestingly, this author blames the rhetorical style of advocates as well as the performance of actors for having tainted the former purity of church preaching. He mentions St Gregory of Nazianzus's aversion to being considered an eloquent and agreeable minister of the Word, and dismissively describes a style of delivery 'qu'ils [preachers] ont fait venir du barreau dans la sanctuaire, fait passer du théâtre à la chaire; de manière, si on l'ose dire, qu'il y a aujourd'hui dans le monde deux sortes de scènes: l'une au théâtre, l'autre dans l'église.'[8] Michel Le Faucheur's *Traité de l'action de l'orateur, ou de la prononciation et du geste* (1657), which is intended for prospective preachers and advocates, observes that some regard a concern with *actio* as unworthy of the priest, believing that religion should deal with the spiritual and intangible (pp. 16–17). Le Faucheur himself does not subscribe to this view, however. Further on he notes the widespread opinion that the advocate has no need of bodily eloquence, because his only legitimate goal is the expression of unvarnished truth, free of seductive ornament such as gesture provides; but Le Faucheur reasons that judges do not always see clearly, and need to be made attentive by the advocate's use of physical expression (pp. 30–3). Le Faucheur holds that nature must be aided by art in oratory (p. 43), and that the senses are more powerfully touched by gesture and verbal *pronuntiatio* than by pronunciation alone (p. 187). Interestingly, Grandmesnil, who acted for the Comédie-Française and Théâtre de la République, and was the last actor to be elected to the fine arts section of the Institut de France, had begun his professional life as an advocate in the Paris parlement.[9]

In 1675 René Bary published his *Nouveau Journal de conversations sur toutes les actions des prédicateurs*, in one of whose dialogues Socrates, Eusebius, and others anachronistically discuss the qualities of a good preacher. For Eusebius, his appearance is important: his face must be severe and his appearance majestic, because 'une mine niaise, inno-

---

[8] *Règles de la bonne et solide prédication* (Paris, 1701), p. 217.
[9] See Albert Soubies, *Les Membres de l'Académie des Beaux-Arts depuis la fondation de l'Institut*, 3 vols. (Paris, n.d.), I. 208–11.

cente et douceureuse' neither touches nor persuades.[10] The interlocutors agree that even the great truths of scripture may benefit from the judicious employment of rhetorical effects (p. 11). Like some other writers of his century and the following one, Bary emphasizes the classical notion that the orator must be moved himself if he wishes to move others. However, the example he cites of Philip of Neri, who always left his listeners distraught because 'il disposait de ses larmes' (p. 114), suggests a factitious rather than genuine show of emotion on the speaker's part. Later in the dialogue Socrates complains about the effeteness of contemporary preachers, products of a world given over to socializing. Although only serious men should occupy the pulpit, one often sees instead 'des amateurs de mots de ruelle, de cercle, de théâtre, et pour dire tout en peu de paroles, que des hommes qui pensent plus à nous chatouiller qu'à nous convertir' (pp. 115–16). But Eusebius regards as desirable in the preacher some of the attributes which a later age was to see as essential in the actor:

il faut que son visage soit presque aussi explicatif que son discours, que ses regards soient tantôt effroyables et tantôt doux, que son front soit quelquefois plissé et quelquefois uni, parce que l'air du visage doit aller de pair avec la qualité des sujets . . . il faut que son geste réponde à son sujet, l'on commet un solécisme de la tête, dit un père de l'église, lorsqu'on parle du ciel et qu'on regarde la terre. (pp. 124–5)

Bary's later *Méthode pour bien prononcer un discours et pour le bien animer* (1679), which deals with both the accent and the gesture desirable in preachers and advocates, offers precise observations on *actio*. To convey frankness, for example, the speaker should spread his arms far apart and hold his hands palm outwards, 'parce que la franchise déploie les plis de l'âme et que les mains tournées en dehors marquent ce déploiement.'[11] Here the relation of outward expression to inward state is a purely symbolic one, but when Bary adds that certain passionate states call for the finger to be laid on the stomach 'parce que le cœur [here located in the stomach] est le siège des passions', he is plainly positing a more direct link between emotion and gesture. Some of the attitudes he specifies are even clearer to the non-initiated reader or observer than the last, such as that of the preacher filled with divine ardour who, with furrowed brow, bends his body so that

[10] René Bary, *Nouveau Journal de conversations sur toutes les actions publiques des prédicateurs* (Paris, 1675), p. 2.
[11] René Bary, *Méthode pour bien prononcer un discours et pour le bien animer* (Paris, 1679), pp. 77–8.

it 'parcour[t] fréquemment la chaire' (p. 106). But some deplored such physical activity. The *Réflexions sur l'éloquence des prédicateurs* which Antoine Arnauld published at the end of the seventeenth century was intended as a rebuttal of the view advanced by another writer, Dubois-Goibaud, that eloquence (including bodily eloquence) should be banished from the pulpit. Arnauld writes with conviction about the need for clerics to practise bodily communication. Six years after him, the anonymous author of the *Règles de la bonne et solide prédication* similarly approves animation of gesture and movement in orators of the pulpit, reasoning that the latter will move their audience more powerfully if they employ *actio* judiciously than preachers whose words, however eloquent and elevated, are delivered without such supports (p. 435).

The presence of an audience which hoped to be entertained as well as edified by sermons was accepted by seventeenth- and eighteenth-century writers, however unwillingly, as a part of communal piety. In 1671 Rapin complained in his *Réflexions sur l'usage de l'éloquence de ce temps* that one daily saw young preachers climb into the pulpit as an actor appeared on stage, to play a role, and that friends and relatives were invited with tickets as though to the theatre (pp. 79–80). Prévost's novel *Manon Lescaut* provides an intriguing example of this practice in the first half of the eighteenth century. Manon's presence at a public debating exercise of Des Grieux's is described by the narrator in these terms:

> Le temps arriva auquel je devais soutenir un exercice public dans l'École de Théologie. Je fis prier plusieurs personnes de considération de m'honorer de leur présence. Mon nom fut ainsi répandu dans tous les quartiers de Paris: il alla jusqu'aux oreilles de mon infidèle. Elle ne le reconnut pas avec certitude sous le titre d'abbé; mais un reste de curiosité, ou peut-être quelque repentir de m'avoir trahi (je n'ai jamais pu démêler lequel de ces deux sentiments) lui fit prendre intérêt à un nom si semblable au mien; elle vint en Sorbonne avec quelques autres dames.[12]

The question of theatricality in preaching is raised again by Dinouart in his *L'Éloquence du corps, ou l'action du prédicateur* (1751). According to Dinouart, most young preachers believe themselves to be orators as soon as they become priests, and support their verbal discourse with a theatrical action that often degenerates into comedy,

---

[12] Prévost, *Histoire du Chevalier Des Grieux et de Manon Lescaut*, ed. Frédéric Deloffre and Raymond Picard (Paris, 1965), p. 43.

of a type which pleases the man of the world and scandalizes the Christian (p. xvi). Later on Dinouart elaborates the notion that the theatre has caused worldliness to infect priestly life (although other culprits are novels and 'les ouvrages colifichets'), and led clerics to practise an improper flattery of voluptuaries and libertines: 'une voix douce, un geste fin et léger, un air fleuri forment leur action. Vous douteriez si ce sont des hommes qui parlent ou des actrices qui déclament: ils en ont l'air et les manières . . . Voilà . . . des orateurs qui se donnent en spectacle à toute une ville' (pp. 137–8). Although action is necessary in the pulpit, Dinouart observes, clerics cannot allow themselves the freedom that was enjoyed by the pagan orators of antiquity to model themselves on actors: the contemporary theatre is a place unfit for men of the cloth to frequent (p. 3). He remarks, however, that since the maxims of religion are contrary to man's inclinations, they must be presented to him with all the appeal that eloquence allows (p. 8). Action is to speech as daylight to the countryside, which remains charming and delightful during the night, but needs the sun for all its beauties and variegated colours to be revealed (p. 14). And to be fully effective, the *actio* of the preacher must be a comprehensive application of all the means at his disposal. Sublimity of action consists in enhancing through facial expression, voice, and gesture the grandeur of one's thoughts, the nobility of one's feelings, and the splendour of one's words (p. 32). In this connection Dinouart later refers to the notion of 'number' in action, which he defines as the harmony resulting from a conformity of bodily movements, facial expression, gestures of the hand, and sounds of the voice with the nature of the matter being discussed (p. 78). But a cautionary note is again sounded as he returns to the comparison with actors. Whereas the latter may perform a role without being convinced by the truth of the words they utter, a preacher's oratory must be a sincere expression of his beliefs (p. 40). Dinouart writes that a different kind of propriety is fitting in the pulpit from that which governs stage performance (p. 79). Dignity is essential in the preacher, while it may be necessarily absent from certain kinds of drama; and 'Loin de la chaire ces orateurs nains, qui ont l'air et la voix d'une marionnette; ils figureraient mieux dans quelque estampe de Callot' (p. 219). Finally, the preacher should remember to show moderation in everything: 'Ne soyez ni manœuvre ni comédien: n'ayez ni trop, ni trop peu de geste' (p. 249).[13]

[13] The actor Talma allegedly influenced preaching style. Regnault-Warin's

Trublet's *Panégyriques des saints, suivis de Réflexions sur l'éloquence en général et sur celle de la chaire en particulier* draws comparisons between judicial and ecclesiastical rhetoric, and distinguishes the separate goals of the advocate and the preacher. The former speaks above all to the mind, and tries to enlighten it, sometimes to seduce the listener's intellect, and always to convince it. But the preacher addresses the feelings and heart, and tries to persuade his audience.[14] The advocate's aim is to be 'raisonneur', and the preacher's to be 'pathétique' (p. 322). Trublet also alludes to the art of acting, and quotes the Jesuit Père Cerutti's warning against the preacher's tendency to 'donner . . . une scène en donnant un sermon'. He must instruct, not amuse; touch with his reasoning, not seduce by his acting; be an apostle, not an acrobat; preach Christianity in a church, not perform it in a theatre (p. 284, footnote).

Nor do warnings of this kind cease to be heard after the Revolution, when the Church has been stripped of many of its powers and lost a part of its moral influence. Just as the Jesuit method of education continued long after the suppression of the Society of Jesus, so fulminations against actorliness in preachers sound on well into the nineteenth century. Maury's *Essai sur l'éloquence de la chaire* (1810), a classic of its kind, describes an essential difference between theatrical action and action in the pulpit. The stage demands contrast in dialogue and gesture which 'le monologue et l'espace de la tribune sacrée ne sauraient admettre . . . Rien n'est donc de plus mauvais goût et plus contraire au ton de la chaire qu'une manière théâtrale.'[15] As an illustration of this truth Maury recalls a 'deplorable' reading by the actor Lekain of the great Condé's funeral oration. Lekain realized after a time, he writes, that the action of a preacher should be less turbulent, although not less animated, than that of an actor in dramatic declamation (pp. 276–7).

Yet the suspicion remained with some commentators that the greatest of French preachers had owed a part of their success and their zeal for preaching to an actorly sensibility. In his *Cours de déclamation* (1804) the actor Larive suggests that Bourdaloue, Bossuet, Massillon,

---

Mémoires de Talma (Paris, 1804), p. 252, reports that a seminarist of St-Sulpice consulted him for advice about inflections of voice and the eloquent use of gesture.

[14] Trublet, *Panégyriques des saints, suivis de réflexions sur l'éloquence en général et sur celle de la chaire en particulier*, 2nd edition, 2 vols. (Paris, 1764), pp. 321–2. See also Sanlecque's *Poème sur les mauvais gestes* (contained in Dinouart, op. cit.), p. 444.

[15] Charles S. Maury, *Essai sur l'éloquence de la chaire*, new edition, 3 vols. (Paris, 1827), p. 276.

and others of their stature enjoyed seeing the emotions aroused in a congregation by their preaching in much the same way as an actor is stimulated by the spectacle of the passions he has awakened in an audience: 'si la religion elle-même nous a donné les Bourdaloue, les Bossuet, les Massillon, etc., etc., nous ne les devons peut-être qu'au charme qu'ils trouvaient à réciter en public leurs éloquents ouvrages, à être témoins des émotions profondes qu'ils communiquaient à un auditoire nombreux, et des larmes qu'ils faisaient verser'.[16]

Boïeldieu remarks in 1804 that men of his day are coldly indifferent to what preachers say, but that nothing brought by the drama of Revolution, in political assembly or on the stage, can match the compelling quality of a preacher's eloquence when he transmits the great truths of religion: 'quel drame sur la scène, quelle action dans les tribunaux peut jamais offrir un intérêt aussi réel et aussi puissant que celui qui naturellement doit naître du discours d'un ministre qui parle au nom même du maître de la nature?'[17] Marmontel, writing before the Revolution, and lacking the standards for comparison which Boïeldieu employs, observes in an article on the 'orateur' that politics provides the would-be orator with the material for his eloquence in the only European country (Britain) in which republican traditions are alive as they were in ancient Greece and Rome.[18] Elsewhere the orator must apply himself to the study of religious morality (for use in the pulpit) or civil and natural law (for use in the courtroom). In Athens the tribune was a kind of theatre in which the great affairs of state were debated, and *actio*, or bodily communication, was a part of performance there; but for moderns such aids to eloquence are inappropriate. Yet the change wrought by the events of 1789 was to create opportunities for the aspiring politician unknown under the ancien régime, and the Revolutionary assemblies opened a field for public eloquence which many commentators compared with the ancient

[16] J. Mauduit-Larive, *Cours de déclamation* (Paris, an XII [1804]), pp. 326–7.

[17] Marie-Jacques-Amand Boïeldieu, *De l'influence de la chaire, du théâtre et du barreau dans la société civile, et de l'importance de leur rétablissement sur des bases qui puissent relever en France leur ancienne et véritable splendeur. Ouvrage politique et moral* (Paris, an XII/1804), pp. 5–6.

[18] Marmontel, *Éléments de littérature*, III. 70, article 'orateur'. On the essential nature of political oratory for the ancient Greeks and Romans, see also Dinouart, p. 13. In *Émile* Rousseau describes the prodigious use made of visual effects in the ancients' eloquence: men *showed* rather than spoke. Of the Romans Rousseau remarks that 'tout chez eux était appareil, représentation, cérémonie' (*Émile ou De l'éducation* (Paris, 1964), pp. 399–400).

forum, and also to the theatre. It is worth examining this phenomenon in some detail.[19]

The advent of the Revolution led a number of observers to relate the political action of the day to the unfolding of a stage drama. Journalists and memoir-writers of the period, as well as speakers in the assemblies themselves, likened current circumstances to tragedy, comedy, farce, or melodrama, according to their own political persuasion. Many députés were, or had been, professionally involved in the theatre. Collot d'Herbois had been an actor and at one time the director of a troupe of actors in Avignon, where according to one witness he found little favour with the public. The author of the *Anecdotes curieuses et peu connues sur différents personnages qui ont joué un rôle dans la Révolution* notes that he was 'sifflé sur [beaucoup] de théâtres de province [et avait] joué la comédie à Lyon, où il était hué, conspué et sifflé régulièrement trois fois par semaine, parce que le théâtre n'ouvrait que trois fois.'[20] Mallet du Pan saw him as resembling the 'tyrans de fantaisie' created by dramatists and bodied forth by Collot himself during his acting career: 'Conspirateur sombre, déclamateur étudié, impopulaire par goût et par habitude, il n'a jamais perdu l'apprêt théâtral.'[21] Collot also wrote plays, all after 1789 being of a patriotic and nationalistic tenor. Fabre d'Églantine, too, was both playwright and actor, although the *Anecdotes curieuses* observes that he achieved little success in either field. Fabre's presumption in daring to appear on the dramatic stage was, according to the anonymous author of this work, astonishing: 'Avec les talents et le moral du comédien Fleury, par exemple, de Préville et de plusieurs de leurs collègues, on est certainement un citoyen précieux et recommandable, mais avec la médiocrité (comme comédien), l'immoralité et la scélératesse de d'Églantine, on n'est qu'un vil histrion' (p. 61). And Mercier writes in *Le Nouveau Paris*: 'j'ai vu Poultier, joueur de gobelets, stentor de spectacles forains, acteur chez le grimacier, même auteur, puis représentant du peuple, et pour couronner tant de gloire, journaliste, et l'ami des lois!'[22]

---

[19] For a more extended account see my article 'The Dramatising of Politics: Theatricality and the Revolutionary Assemblies', *FMLS*, XX (1984), 193–212.

[20] *Anecdotes curieuses et peu connues sur différents personnages qui ont joué un rôle dans la Révolution* (Geneva, 1793), p. 34.

[21] Mallet du Pan, *Mémoires et correspondance*, ed. A. Sayous, 2 vols. (Paris, 1851), II. 45.

[22] L.-S. Mercier, *Paris pendant la Révolution, ou Le Nouveau Paris*, 2 vols. (Paris, 1862), II. 447.

In *Les Origines de la France contemporaine* Taine mentions actors as having been among the group of men ambitious of a position in government after the events of 1789 (others he names are advocates, brochure-writers, and journalists).[23] It was certainly widely acknowledged that actors had developed through their art some of the aids to eloquence which could help orators of the tribune to make their mark in the Assemblée. They gave advice, for example, to the advocate Hérault de Séchelles[24], who became an important political figure in the Revolution. In this connection Lemercier declares in his tract *Du second théâtre français, ou instruction relative à la déclamation dramatique* (1818) that since perfect dramatic delivery consists in every variety of declamation, the teacher, advocate, député, and all the organs of democratic government should attend the theatre as a school for language, taste, eloquence, poetry, and the passions of men. Sometimes the story was told of Garrick's visit to the House of Commons, in the course of which Fox declared the whole chamber to be indebted to him for lessons in the art of declamation.[25]

Something of the theatricality which came to be associated with proceedings at the Assemblée and Convention nationale had existed before in France. Attendance at sessions of the parlements, especially that of Paris, was regarded as in some ways similar to visiting the theatre. Dissident speakers were liable to be booed, and although the assemblies were in theory private, small boxes or 'lanternes' were available for privileged spectators, who could thus watch proceedings secretly.[26] After the creation of a national assembly in 1789, references to its sessions as a form of theatre abound. Marmontel notes in his memoirs that the rooms at Versailles destined for general assemblies, in which the most important affairs of state would be debated between the three estates, were surrounded by galleries as though to invite the people to be present at debates, support one party, vilify the opposition, and transform the tribune into a stage where actors earned applause.[27] But Necker, he continues, had envisaged such assemblies

[23] Hippolyte Taine, *Les Origines de la France contemporaine*, 6 vols. (Paris, 1876–94), II. 116.
[24] After Huerne de la Mothe was struck off the register of advocates for his attempt to prove the injustice of excommunicating actors (in *Libertés de la France contre le pouvoir arbitraire de l'excommunication*, 1761), he joined a theatre troupe.
[25] *Journal des théâtres, de la littérature et des arts*, 24 pluviôse an VII (12 February 1799), p. 298.
[26] Charles Aubertin, *L'Éloquence politique et parlementaire en France avant 1789* (Paris, 1882), pp. 186–9.
[27] Marmontel, *Mémoires*, ed. John Renwick, 2 vols. (Clermont-Ferrand, 1972), II. 356.

as being 'un spectacle paisible, imposant, solennel, auguste, dont le peuple aurait à jouir' (ibid.). Both at Versailles and, later, in Paris the Assemblée met in vast halls capable of holding two thousand people, which obliged speakers to shout and gesticulate in order to be understood. Arthur Young observes that during one session he attended over a hundred députés were on their feet simultaneously, each trying to win an audience (Taine, op. cit., pp. 144-5). The speaker's art of *actio* gave him a clear advantage here.

Furthermore, many writers of the day claimed that the Assemblée had taken over from regular theatres as a source of popular dramatic entertainment. The author of the tract *De l'organisation des spectacles de Paris* (1790) observes with reference to the fortunes of the Comédie-Française, or Théâtre de la Nation, that the times are not propitious for the company. Many of their old patrons (drawn from the upper strata of society) have emigrated, and foreigners no longer visit Paris. Furthermore, the continual assemblies of commune and district offer the idle among the city's population an interesting and free variety of spectacle which is a very acceptable substitute for the theatre.[28] (In the early days of the Assemblée nationale, the visitors' gallery was largely filled with the well-to-do; only later were they displaced by a 'popular' crowd of a type which the Comédie-Française had never catered for.)[29] The actor Fleury confirms the substance of the anonymous author's assertion in his memoirs, where he suggests that the detrimental effect of the Revolution (as he sees it) on dramatic spectacles may have been brought about by the continual diversion of national interests, which replaced the people's usual amusements, and also by the emigration of so many Parisians, especially the wealthy.[30] Although Fleury refers here to 'tous les spectacles', his argument seems applicable to the Comédie-Française rather than to playhouses which had traditionally purveyed 'popular' spectacles, or to the recently-founded Théâtre de la République, whose repertoire was more in tune with the current preoccupations of republican Frenchmen.

An undated brochure attributed to Mirabeau, *La Lanterne magique nationale* (which internal evidence shows to have been written after 1789), describes a series of events in Versailles and Paris as though they were being projected by a magic lantern showing images of

---

[28] *De l'organisation des spectacles de Paris* (Paris, 1790), p. 67.
[29] See Roger Garaudy, *Les Orateurs de la Révolution française* (Toulouse, 1939), p. 20.
[30] *Mémoires de Fleury de la Comédie-Française*, 6 vols. (Paris, 1835-8), IV. 305.

politically significant scenes, and in so doing adds emphasis to the notion that political business of the time was commonly regarded as a form of visual as well as verbal entertainment. One picture reveals the 'grandes marionnettes à la salle du tiers', a 'spectacle de nouvelle création'.[31] The reference, clearly, is to the convening of the States-General in May and June of 1789, in which the three orders occupied separate rooms. Two thousand spectators are lodged around the perimeter of the hall. The presenter remarks that he would, if he had the means, make the 'orateurs' speak, and 'je les ferais, j'espère, mieux parler qu'ils n'ont fait.' A later slide, for 23 June 1789 (the date on which Louis XVI tried in vain to annul the decision of the 'tiers état' to form a national assembly), shows the king surrounded by his family and ministers, with 'le grand génie de la finance'—Necker, who despite aristocratic opposition had doubled the number of representatives from the 'tiers'—conspicuously absent. He organized the show and directs it, the author tells us; but puppets deserve our applause only when we cannot see their strings. Necker is behind the curtain, and not unless the play succeeds will he admit that he wrote it.[32] In a sequel called *La Nouvelle Lanterne magique*, as though to demonstrate that politics and public entertainment are of a piece, the scene shown is that of the fête de la Fédération on 14 July 1790, called the greatest spectacle ever seen.

Barère's memoirs give a further illustration of the tendency to view political proceedings in theatrical terms. Barère recalls hearing an eloquent speech delivered in January 1789 at a public session of the États de Languedoc, which attracted a number of foreigners. An Irish archbishop, Dillon, spoke with the kind of eloquence that connoisseurs admired in Fox. Barère continues:

Je me trouvais à côté du célèbre acteur comique M. Préville, qui écouta avec une attention toute particulière ce genre d'esprit et d'éloquence dont les chefs-d'œuvre du théâtre ne lui avaient pas donné une idée . . . [At the end, Préville gives Barère his opinion of the speaker.] Ma foi, nous dit-il, il a supérieurement joué son rôle . . . Tant il est vrai que ces grands acteurs ne voient partout qu'une comédie et des comédiens.[33]

The name of Mirabeau was often linked with the dramatizing of politics. Fleury recounts the story of his colleague Molé's visit to the

---

[31] [Mirabeau,] *La Lanterne magique nationale* (n.p., n.d.), p. 16.
[32] On Mirabeau's dislike of Necker, see G. Chaussinand-Nogaret, *Mirabeau* (Paris, 1982), p. 128.
[33] B. Barère de Vieuzac, *Mémoires*, 4 vols. (1842–4), I. 238.

Assemblée on a day when Mirabeau's eloquence, gestural as well as verbal, shone with especial brilliance, and where Molé was struck by

la voix, les gestes, l'organe de Mirabeau, ce feu, cette phrase abondante, cette grande et terrible image du gouffre entr'ouvert, cette main qui, raidie et effrayée, se portait en avant pour désigner les profondeurs de l'abîme, tandis que l'autre s'attachait à la tribune avec l'action d'un homme qui saisit sa planche de salut. (IV. 167)

Mirabeau's 'entraînante action oratoire' astonished and dumbfounded Molé, who declared to the député when he had descended from the tribune, 'Ah! monsieur le Comte . . . quel discours! quelle voix! quels gestes! mon Dieu! mon Dieu! que vous avez manqué votre vocation!' (ibid.)[34] Appropriately enough, the occasion for this display of Mirabeau's was the discussion whether Chénier's tragedy *Charles IX* should be performed as part of the fête de la Fédération or not. Dumont's *Souvenirs sur Mirabeau* also reports this memorable scene, and interestingly reveals that Mirabeau's speech, like a great many of his others, was a 'ghosted' one.[35] The great orator, in other words, resembled the actor in that he was not the author of the discourse he delivered.

Desmoulins recalls in number XXVIII of his *Révolutions de France et de Brabant* that Mirabeau's performance in the assembly was in this respect like that of Roman actors who formed pairs for the playing of parts, one to speak and the other to provide the action (pp. 184–5). And Du Roveray, an aide of Mirabeau's, described him as a 'mannequin auquel nous avons fait jouer si longtemps un beau rôle' (Dumont, p. 36). But the *Mercure* for September 1792 affirms that he was never so great as when he spoke spontaneously (p. 82). His gesture, facial expression, and 'toute son action extérieure' shook and then exalted his entire audience, and 'la chaleur [de ses] mouvements . . . précipitaient ses phrases' (ibid., p. 74). Mirabeau's dislike of facile effects is noted by Dumont, who writes that the orator found an 'operatic' style of declamation distasteful (p. 157). Many commen-

---

[34] De Ferrières notes: 'Mirabeau joignait aux talents naturels, qui font les orateurs, une étude réfléchie de l'art oratoire. Il savait que l'homme de génie parle encore plus aux sens qu'il ne parle à l'esprit: aussi son geste, son regard, le son de sa voix, tout, jusqu'à la manière de se mettre et d'arranger ses cheveux, était calculé sur une connaissance approfondie du cœur humain.' ([C. E. de Ferrières,] *Mémoires pour servir à l'histoire de l'Assemblée constituante et de la Révolution de 1789*, 3 vols. (Paris, an VII), I. 89.)

[35] Étienne Dumont, *Souvenirs sur Mirabeau et sur les deux premières Assemblées législatives*, ed. J. Bénétruy (Paris, 1951), p. 120.

tators remark on the impressiveness of his *actio*. Touchard-Lafosse writes that he was superior to the Girondin Vergniaud, even though the latter's 'geste' was neither studied nor theatrical, and his bodily movements, supple and precise, added to the seductiveness of his verbal delivery.[36]

The anonymous author of the pamphlet *Mirabeau à l'Assemblée constituante* observes that a contemporary described Mirabeau's gestures as just and pronounced, and wrote that at the tribune his manner was noble, and his eyes were full of fire.[37] According to Girardin, Boze's capturing of their brilliance made his portrait of Mirabeau bearable, even though the artist had also depicted all the man's celebrated ugliness.[38] Nearly every commentator mentions Mirabeau's physical unattractiveness, but most conclude that it was no obstacle to his achieving success in the Assemblée. (A parallel with Lekain comes to mind here.)[39] In the view of Népomucène Lemercier, when he was speaking his eloquence conferred grandeur on his whole being. He acquired true beauty, and his very vigour was graceful, utterly transformed by his soul.[40] The same writer remarks that Mirabeau's inward state governed his gestures, which were pronounced but rare; it appeared in his dignified gait and proud bearing; and 'son génie accordait noblement et sans grimace le feu de ses regards, le tressaillement des muscles de son front, de sa face émue et pantelante, et le mouvement de ses lèvres, aux intonations de la vérité, de la véhémence, de la menace et de l'ironie' (ibid.). People who had seen Mirabeau and Lekain together, he continues, assured him that they might have been mistaken for one another, 'aux conformités de leur naturel également théâtral' (p. 38). Mérilhou's *Essai historique sur la vie et les ouvrages de Mirabeau* (1827) observes that Mirabeau's gestures were 'rares, mais justes, nobles, prononcés, véhéments quelquefois, et toujours en harmonie avec les intonations du discours'.[41] Desmoulins, on the other hand, is struck by the theatricality of his action

---

[36] Georges Touchard-Lafosse, *Histoire parlementaire et vie intime de Vergniaud, chef des Girondins* (Paris, 1848), p. 20.
[37] *Mirabeau à l'Assemblée Constituante* (Paris, 1848), p. 5.
[38] Stanislas Girardin, *Mémoires et souvenirs*, 2nd edition, 2 vols. (Paris, 1829), I. 113.
[39] Lacretelle writes that 'Par-dessus tout, il était un élève du célèbre acteur Lekain. Il savait tirer parti de ses défauts mêmes. Sa laideur avait une expression tragique, et disparaissait quand il était sublime.' (Charles Lacretelle, *Dix Années d'épreuves pendant la Révolution* (Paris, 1842), p. 36.)
[40] Népomucène Lemercier, *Du second théâtre français, ou Instruction relative à la déclamation dramatique* (Paris, 1818), p. 37.
[41] Mérilhou, *Essai historique sur la vie et les ouvrages de Mirabeau* (Paris, 1827), p. ccx.

on occasion, and describes a point during a session at the Assemblée when Mirabeau resorted to extravagant movement after other means had failed to command quiet for him to be heard. Seeing that he could not obtain a hearing, he decided on a visual form of address, and clambered on to a seat instead of donning his hat (*Révolutions de France et de Brabant*, LV. 113). But even this action, according to Desmoulins, failed to attract attention.

Mirabeau was a friend of Talma, who admired him greatly. After Mirabeau's death Talma had the following inscription engraved on the door of the orator's house in the rue Chaussée d'Antin:

> L'âme de Mirabeau s'exhala dans ces lieux;
> Hommes libres, pleurez; tyrans, baissez les yeux.[42]
> (*See Plate 3*)

The bishop of Autun reported to Dumont apropos of this death that after the final attack of the illness from which he died, Mirabeau regained all serenity, and dramatized his last moments: 'il se voyait l'objet de l'attention générale, et il n'a cessé de se parler et de se conduire comme un grand et noble acteur sur le théâtre national' (*Souvenirs sur Mirabeau*, p. 170).

Much concern is expressed from the end of the sixteenth to the end of the eighteenth centuries about the status of particular kinds of rhetoric. In different ways, the eloquence of the preacher and of the man holding or desiring to hold political office are seen to be restricted, in the first case because of concern about the propriety of rhetorical aids, and in the second because the scope for exercising eloquence is limited by the prevailing political circumstances. The possibility that the accompaniment to verbal eloquence of physical action may demean the subject of the speaker's discourse by appealing to the senses of the audience as well as to its intellect is entertained, to varying degrees, by practitioners of ecclesiastical, judicial, and political eloquence, and the need for sobriety in their cases often contrasted with its absence in the actor's.[43] *Actio* is believed to enhance

---

[42] Émile Duval, *Talma. Précis historique sur sa vie, ses derniers moments et sa mort* (Paris, 1826), pp. 79–80. Duval also notes of Talma that 'Tant qu'il fut témoin des débats furieux de nos tribuns populaires, il ne chercha point à retenir l'essor de sa fougue; alors il fut Brutus, plus tard il fut César' (p. 116). On the attendance of prominent republicans at Julie Talma's salon, and her favouring of the Girondins, see Alfred Copin, *Talma et la Révolution* (Paris, 1887), p. 100.

[43] I shall later discuss the disapproval of exaggerated *actio* in performance which actors themselves expressed.

the immediacy and hence the potency of the orator's words, and yet some writers declare that too much enhancement or artistry is undesirable. The latter opinion is related to a general suspicion of artistry in rhetoric which goes back to antiquity. In the eyes of some classical commentators and their successors, eloquence is to be employed with caution because the ornateness and 'show' which may be its concomitant are diversions or obfuscations which deflect the listener, by seductive means, from a possibly unpalatable truth. This is not the place to discuss the suspicion of ornate and potentially unreliable eloquence in detail. It is enough to note its connection with the unease felt about the falseness or imitativeness of art from Plato onwards. The untruthfulness of conventional drama became a subject for discussion in the eighteenth century by realist playwrights like Diderot, but such untruthfulness was at least as strongly criticized by contemporaries in the case of more traditionally sanctioned fields of eloquence, particularly where, as in pulpit oratory, a moral need for reliability was perceived.

More than one writer on the rhetoric of the Bar observes that it is of less concern that the advocate should be expressing the unvarnished truth than that he should persuade his audience—whether judge or jury—by the force of his arguments, but that the same is not true where religious doctrine is the speaker's subject. The art of drama, it may be felt, is closer in this respect to judicial pleading than to preaching; and this was precisely the complaint addressed to it by some commentators, most notably Rousseau in the *Lettre à M. d'Alembert sur les spectacles*. The great artist, he claims in his discussion of *Phèdre*, can persuade his audience to put its normal impulses to moral judgement in abeyance, and elicit a response (such as the sympathy felt for Phèdre) which seems valid in a particular case, but which lacks permanent applicability. The proper moral sentiment is thus subverted. Interestingly, this question is raised for modern critics by the theory and practice of *drame* playwrights in eighteenth-century France, but for the opposite reason. In the *drame*, it is generally agreed, moral issues are (unartistically) presented with too much clarity, and right action given too unequivocal a preponderance over wrong. As a result, the audience often finds the alleged moral dilemmas of the characters uninteresting.

Rousseau's argument in the *Lettre* is part of the general mistrust of art which he had expressed much earlier, in the *Discours sur les sciences et les arts* of 1750. He does not discuss the matter which concerns me here, namely the actor's reinforcement of a drama's verbal eloquence by means of bodily movement and attitude. But many of his contemporaries or near-contemporaries do engage in such a discussion, whether they relate their observations specifically to dramatic *actio* or treat drama in the light of general beliefs about the value of gesture and movement in the arts of public speech.

Cicero's very vague definition of *actio*—'est enim actio quasi corporis quaedam eloquentia, cum constet e voce atque motu' (*Orator*, xvii. 55)—was often repeated without much further investigation of its subject in eighteenth-century discussions. But in the course of this century, and on into the nineteenth, there was an increasing interest in the possibility of constructing a science of *actio* which would reduce to rule the various movements appropriate to different circumstances, and thus facilitate the teaching of this branch of rhetoric. Initially such a system was envisaged as applicable to aspiring preachers, advocates, or statesmen—members of the liberal professions for which the colleges prepared pupils—but as the acting profession acquired sufficient respectability for it to be endowed with schools in which young performers were trained, its principles would in theory be available to teachers in those institutions too.[44]

The advocate and député Hérault de Séchelles, writing about the art of declamation (which he glosses as the ancients' *actio*),[45] expresses the opinion that the fundamental principles of effective bodily action are the same for performers at the Bar and in the theatre. He confirms that he himself learned much from watching actors (p. 416), and even took lessons from Mlle Clairon (p. 403). Dinouart's manual on *L'Éloquence du corps, ou l'action du prédicateur* gives a number of very general instructions on the subject, but also offers precise rules. Many of them, he implies, are derived from what the ancients had written on the subject (an interesting illustration of the way in which moderns drew on the work of non-Christian predecessors), and from the teachings of his seventeenth- and eighteenth-century forebears. The work of seventeenth-century English writers, although rarely mentioned by French commentators, had made an important contribution to precepts on bodily communication. John Bulwer, who was also known for his efforts at educating the deaf and dumb, published in 1644 a *Chirologia* and *Chironomia* which offered specific instructions

[44] See chapter 6.
[45] Hérault de Séchelles, 'Réflexions sur la déclamation', *Magazin encyclopédique, ou Journal des lettres, des sciences et des arts*, I (Paris, 1795), 396–416 (p. 396).

on the use of the hands and arms in rhetorical delivery. In 1604 Wright had noted that orators and actors were largely agreed on the substance of external action.[46] Elizabethan acting was designed to express the spirit through bodily attitude and movement, and like French theorists of the late seventeenth and the eighteenth centuries, the Elizabethans saw body and soul as aspects of one another. *Actio*, for them, showed through gesture a state of physiological secretion in the body as well as the condition of the soul (p. 100).

Some of Dinouart's observations concerning bodily carriage, movement of the head and shoulders, and facial expression (p. 227 ff.) are clearly based on Lebrun's celebrated typology as expounded in his Academy lectures and accompanying illustrations of 'têtes des passions'.[47] The last point applies equally to Dubroca's *Principes raisonnés sur l'art de lire à haute voix, suivis de leur application particulière à la lecture des ouvrages d'éloquence et de poésie*. This work is intended in the first instance for those who read their works out in public, but addresses itself in fact to all who try through verbal and other kinds of eloquence to 'Toucher, éclairer, convaincre, instruire, émouvoir, ou amuser' (p. vii). With this in mind, Dubroca writes at some length about the 'jeu de la physionomie, de l'attitude du corps, des règles de bienséance, et des gestes qui doivent accompagner une lecture'. After dealing in his first three sections with the methods of touching the heart, convincing the mind, and captivating the ear, Dubroca turns to the 'moyens de plaire aux yeux, ou de l'action extérieure du lecteur', and notes that the rules for bodily expression are equally useful to orators (what kind is not specified), professional actors, public speakers, and those who want simply to read out loud effectively (p. 356).

He discusses external action with reference to three areas of expression: bodily attitude, the play of features, and gesture. Some types of performance, he notes, do not require much modification of the countenance (p. 361). Such are the reading aloud of historical and philosophical works. But 'ouvrages de sentiment', and especially public speeches intended to produce a great effect, require that the speaker give his countenance the tone of the matter being aired. On the subject of the eyes, he makes specific reference to the theatre:

[46] See B. L. Joseph, *Elizabethan Acting* (Oxford, 1950), pp. 2–3.
[47] On these matters see also Dene Barnett, 'The Performance Practice of Acting: The Eighteenth Century', *Theatre Research International*, new series, II (1976–7), 157–86; III (1977–8), 1–19, 79–93; V (1979–80), 1–36; VI (1980–1), 1–32.

'l'expression des yeux et du visage est l'âme de la déclamation théâtrale; c'est là que les passions doivent se peindre en caractères de feu' (p. 518). The ancients, who wore masks, had no idea of this type of expression; but moderns have the advantage over them of unmasked faces and small theatres (ibid.). The art of painting is seen by Dubroca as exemplifying the clear expression of feeling through facial expression: 'Pourquoi est-il si facile aux hommes même les moins judicieux de deviner quel sentiment anime chacun des personnages d'un grand tableau? Parce que le peintre a donné à la contenance, au maintien de chacun d'eux le caractère qui lui convient' (p. 363). Dubroca does not pause to consider whether similar gestures are appropriate to arts in which *actio* is combined with words, and where the depiction of action in time as well as in space is possible. He draws other examples from the visual arts, pointing to the paintings of Lebrun, Le Sueur, Poussin, and David, 'où toutes les figures sont des espèces de pantomimes, d'autant plus admirables que pour s'exprimer, elles ont, non une suite de gestes qui s'entr'aident réciproquement, mais seulement un geste qui est unique' (p. 384).

The question of nature and art, in *actio* as well as in other kinds of declamation, is recurrently raised by eighteenth-century theorists of acting. Cailhava's treatise on the decadence of theatre refers back to the seventeenth-century acting style, and describes the excess that was its hallmark before Molière set about reform. He speaks of 'ces convulsions, ces tortillements de bras, cette déclamation chantante, ce jeu forcé, précieux ou *taquin*, cette monotonie assommante qui régnait au théâtre si tyranniquement lorsque Molière, fléau de tous ces vices, parut et les détruisit.'[48] Many contemporaries, none the less, spoke of the importance of gesture and action in Molière's own performing style.[49] In the eighteenth century, Luigi Riccoboni writes that a degree of enlargement beyond nature is acceptable in the performance of drama as well as in preaching and political debate, because many members of the audience will be distant from the speaker. But still, Riccoboni stresses, exaggeration should be accompanied by circumspection (pp. 29–30). Since moderation is a relative quality, it is not surprising to find some disagreement between commentators as to its desirability in acting. Certainly, as we have seen,

---

[48] Cailhava de l'Estendoux, *Les Causes de la décadence du théâtre* (Paris, 1807), p. 11.
[49] See Roger Herzel, 'Le Jeu "naturel" de Molière et de sa troupe', *XVIIe Siècle*, 132 (1981), 279–83.

this art is often contrasted to other forms of public address with respect to the need for control. A late theorist, the Englishman Austin, remarks that while political orators must constantly keep their emotions in check in order to mould the opinions of others, the same degree of self-government is unnecessary in actors (pp. 241–2). They may represent characters strongly, and are limited only by the boundaries of decency. Mlle Clairon describes Mlle Dumesnil's acting style as sometimes excessively gestural, although she admits a saving grace in her rival: 'Ses gestes étaient souvent trop forts pour une femme; ils n'avaient ni rondeur, ni moëlleux; mais ils étaient au moins peu fréquents.'[50] Talma, on the other hand, approves Mlle Dumesnil's naturalness in this respect, and like Diderot before him criticizes Mlle Clairon's performing style as having been 'tout d'arrangement et de calcul'.[51]

D'Hannetaire, himself a former actor, similarly recommends the 'rule of nature', and implies that dramatic performances should approximate as closely as possible to the conduct of ordinary life. He sees the ideal state for the actor as being one of absorption in his role, to the extent that he forgets he is acting. The result is an unconsciousness of gesture which is similar to the unstudiedness of ordinary social intercourse: one can scarcely say at the end of a lively and interesting conversation whether one's gestures as one spoke were good or bad, and this is because they were natural and spontaneous.[52] Despite the grandeur that tragedy conventionally demands of its protagonists, tragic gestures, according to this writer, should remain moderate and in nature. Again, d'Hannetaire recommends self-forgetfulness in the actor, for this will result in performance consonant with the part being played, and bodily eloquence will match verbal expression. If the performer is capable of immersing himself in a role, that is if he has a feeling heart, the right gesture will inevitably follow, and it is positively dangerous to think about it too much (ibid.). D'Hannetaire

[50] Hippolyte Clairon, *Mémoires*, new edition (Paris, 1822), p. 290. Geoffroy denies this, however: 'Mlle Dumesnil était avare des gestes, disait avec beaucoup de simplicité, quelquefois même de négligence.' (Julien-Louis Geoffroy, *Manuel dramatique à l'usage des auteurs et des acteurs* (Paris, 1822), pp. 224–5.)

[51] François Talma, 'Réflexions sur Lekain et sur l'art théâtral', in Henri-Louis Lekain, *Mémoires* (Paris, 1825), p. x. See also Diderot, *Le Neveu de Rameau*, ed. Jean Fabre (Geneva, 1963), p. 54. Subsequent references to *Le Neveu de Rameau* are to this edition.

[52] J.-N. Servandoni d'Hannetaire, *Observations sur l'art du comédien et sur d'autres objets concernant cette profession en général, avec quelques extraits de différents auteurs et des remarques analogues au même sujet*, 2nd edition (Paris, 1774), p. 205.

writes that there is a golden mean in the enactment of *geste*. The actor should no more remain stiff than he should flail with his arms and appear convulsed (p. 206). This recommendation is very close to the advice which contemporary rhetoricians give to priests and other public speakers. It parallels the observations in Sanlecque's satirical *Poème sur les mauvais gestes*, in which the author counsels the preacher to conduct himself with moderation in the pulpit:

> Du geste et du sens la mesure pareille
> Doit autant charmer l'œil qu'elle charme l'oreille;
> Si le geste et le sens sont toujours de complot,
> Un seul geste jamais ne dément un seul mot.[53]

Grand opera was often seen as calling forth grandiloquent gesticulation from its performers, despite the fact that eighteenth-century writers on acting often declared their observations to be as applicable to the former art as to non-lyrical drama. Favart, for instance, observes in a letter of 6 February 1763 that opera encourages a non-natural acting style, and remarks of a young actress who was formerly a pupil at the Académie royale de musique, but has not yet appeared on the Opéra stage, that '[elle] n'est point encore faite aux grands gestes de l'Opéra; tant mieux, elle sera plus naturelle.'[54] At the same time, he remarks on another occasion, performers of opera make no effort to combine their singing with effective bodily action. He describes Mlle Rozetti's début as having revealed both her astonishing voice and her complete ignorance of the art of movement: 'Elle ne fait que lever les bras, et les poser ensuite sur son honneur en honnête fille étonnée de se trouver dans un lieu où elle court tant de risques' (I. 84–5). Another singer, Joli, who made his début at the same time, is similarly criticized: '[il] manque encore d'action' (ibid.). Action is indispensable to spectacle of this kind, Favart asserts: only the Italians can manage without it. The French insist that silence itself should be expressive, whereas Italians can arrange their jabots while humming the sound of a tempest (ibid.).[55] Dorat noted in the section on opera of his poem *La Déclamation théâtrale* that the opera singer could not be excused from exhibiting the most expressive mien and gesture,

---

[53] Louis de Sanlecque, *Poème sur les mauvais gestes*, in Dinouart, *L'Éloquence du corps*, pp. 443–4.

[54] C.-S. Favart, *Mémoires et correspondance littéraires, dramatiques et anecdotiques*, 3 vols. (Paris, 1808), II. 65.

[55] See also Dene Barnett, 'La Rhétorique de l'Opéra', *XVIIe Siècle*, 133 (1981), 335–48.

although he observed that Sophie Arnould was 'la seule actrice de l'opéra'. Yet in opera, as Austin later remarked, all interest in the progress of action might be lost because of the difficulty of following the dialogue delivered in recitative (p. 248). Dorat had consequently suggested that true recitative, where most of the action occurs, should be declaimed rather than sung.

Some writers, nevertheless, emphasize the fact that there are moments in theatrical performance when few gestures, if any at all, are required. Dubroca remarks that the despondency of grief virtually prohibits action, and that reflective states cannot allow it. In the theatre, he writes, emotions like indignation, contempt, pride, and fury should be expressed simply through the eyes. Any accompaniment to them in facial expression generally, movement of the head, or gesture would weaken the effect; and when one reproaches a performer with neglecting gesture in acting a pathetic father or a majestic king, one has forgotten the rule that 'dignity has no arms' (p. 517). In *The Analysis of Beauty* Hogarth observes that, just as blank spaces in painting add considerably to the overall beauty of the composition, so there is an absolute necessity in acting for an occasional cessation of movement. There was a great need on the contemporary stage, in his view, for relief from what Shakespeare called 'continually sawing the air' (p. 152).

The prince de Ligne, writing in 1774, remarks that the kind of wild gesturing which marred acting styles before Molière's reforms was still evident a century later. He recalls d'Aubignac's description of Mondory's manner of performing, which was characterized by much agitation and head-shaking,[56] and offers the following observation on the *actio* of some moderns: 'Il n'y a rien de pis . . . que ces énergumènes qui veulent tout peindre, qui ont le geste à la chose, qui jouent le mot, qui se démènent, roulent les yeux terribles, agitent des bras et des jambes, et ressemblent plutot à des convulsionnaires qu'à des comédiens' (p. 120). Fourteen years later Levacher de Charnois looks back with admiration to Baron and Mlle Lecouvreur, who accustomed the public to a manner of performing tragedy that was as noble as it was natural.[57] This did not mean that they failed to convey physically the transports to which tragedy gives rise; but their air of abandon, which was justly prized, was the product of profound emotion, not

[56] [Prince de Ligne,] *Lettres à Eugénie sur les spectacles* (Brussels, 1774), p. 114.
[57] [Jean-Charles Levacher de Charnois,] *Conseils à une jeune actrice* (n.p., 1788), p. 24, note 14.

of a fevered brain. He criticizes Larive for pandering to the contemporary public's taste for excess, which precludes the feeling of pity to which tragedy should move its audience. Hercules, he declares, should not remind us of a criminal on the wheel in his torment. The more he keeps it within himself, betraying it only through inarticulate moans and half-suppressed cries, the more he will impress as a hero (p. 25, note 15). Like Dubroca after him, Levacher states the belief that facial expression may be enough to retain the audience's attention, and be a sufficient complement to the dramatic dialogue. He disapproves of the exaggerated pantomime and constant genuflections which are mistakenly thought to increase interest: the play of the physiognomy combined with clear articulation often suffices (p. 35).

Looking back over a long life, in the course of which he had developed a lasting taste for the theatre and become the most influential drama critic in Europe, Geoffroy reflects in 1822 on the way second-rate actors conceal their want of sensibility by '[les] grimaces, les cris, les attitudes forcées, les gestes multipliés, les prestiges de la pantomime . . . c'est ce qu'on appelle *se battre les flancs*' (pp. 16–17). Sadly, he says, this style of acting will remain, because it impresses the common people, and because nothing is rarer in actors than true sensibility. On a different occasion he attacks the new generation of actors (to which Talma belongs) for abusing gesture, and writes that their insistence on acting 'd'instinct et par une inspiration soudaine' makes them 'les quakers de l'art dramatique'.[58] Of Talma's performance as OEdipe he observes that 'sa figure et son jeu muet, si l'on en excepte les gestes, méritaient des éloges' (ibid., p. 224).

The most complete discussion of these matters, as of much else relative to acting, is to be found in the theatre director Dorfeuille's work *Les Éléments de l'art du comédien, considéré dans chacune des parties qui le composent, à l'usage des élèves et des amateurs du théâtre*. All of his remarks on gesture emphasize the need for it to be controlled and used sparingly. In the actor's 'jeu muet' (or 'jeu de théâtre'), where Dorfeuille allows him to improvise without any direction, the performer proves his sensibility and is at once true and natural; but this type of acting must not be exaggerated or occur too frequently.[59] In his section on gesture, Dorfeuille provides a very broad definition of the word: 'Au théâtre on appelle geste tous mouvements, signes, jeu de la

[58] Julien-Louis Geoffroy, *Cours de littérature dramatique*, 6 vols. (Paris, 1825), VI. 280.
[59] [Dorfeuille, i.e. P.-P. Gobet,] *Les Éléments de l'art du comédien, considéré dans chacune des parties qui le composent, à l'usage des élèves et des amateurs du théâtre*, 9 vols. (Paris, an VII–an IX), III. 31–3.

physionomie, attitudes et expressions muettes' (IV. 4). Dorfeuille's conviction that the actor should always strive for truthfulness and naturalness in his performance leads him to declare that gesture cannot be learnt, and should not even be rehearsed. This is surprising, perhaps, in view of the very detailed instructions about *actio* which his handbook goes on to provide. His fourth book, for instance, furnishes the following precise description of a graceful gesture of supplication: 'les bras s'élèvent à la hauteur de la poitrine et sont écartés d'environ un pied. Si la grâce est personnelle, la tête est inclinée en avant, l'œil est baissé, les mains sont rapprochées et se joignent lorsque la prière devient plus instante; le corps dans cette attitude reste immobile après qu'on a supplié, il en attend l'effet' (ibid., pp. 23–4). Elsewhere the importance of facial expression is stressed; hand movements are described in detail, as they had already been in books on rhetoric; and so on. Dorfeuille's belief that all gesture should be true and natural is formulated in terms which express the distinctiveness of 'truth' in the different artistic mediums. He does not regard the model of the painter's canvas or the sculptor's marble as necessarily suitable for the actor, whose art is one of movement. Thoughtless copying is therefore discouraged: positions imitated from the painter or the sculptor, he writes, are always cold, and incapable of suggesting action effectively (ibid., p. 38).[60] Dorfeuille's ideal for the actor proves to be a combination of nature and application, spontaneity and study, for he declares that the actor should strive to render his gestures agreeable, to give suppleness to his movements and truth to his attitudes (ibid., p. 5). He notes that the abundance of gesture is always a fault, suggesting unsettledness or some other want of substance in the character. He is an enemy of affectation, demanding of the tragedian a weightiness and measure consonant with the seriousness of the subject being treated. 'Le geste doit être partout développé et soutenu, arrondi, grand, sans être jeté ni précipité' (p. 8). More than that, the actor must seek balance in his movements, even when the tragic action has reached the point of disharmony and disintegration, for propriety dictates that geometrical and harmonious proportions be maintained (p. 10). Above all, the actor should remember that gesture is an auxiliary: not a replacement for language, but a means of enhancing it (a point which earlier commentators had also stressed). Gesture should not dispute with voice the right to be remarked first: rather, it should remain the agent of voice, and

[60] See chapter 3.

support rather than supplant it (p. 11). Thus Dorfeuille explicitly distances himself from the view of Mlle Clairon that no rules useful to the actor's art exist (I. 34), and plainly counts *actio* among those aspects of performance in which instruction is both feasible and desirable.

For some informed observers at the end of the eighteenth century, then, the actor's profession unquestionably merited equal consideration to that accorded others involving the art of declamation. The bodily eloquence of the actor, in their opinion, resembled that of more respected performers in church or courtroom: if the degree of *actio* varied from one to another, its essential nature was constant. This being so, it was wholly legitimate in the eyes of commentators like Dorfeuille to posit common principles of instruction for all whose professional activity embraced the art of *actio*. Among earlier writers too, there was broad agreement about the role of bodily expression in the speaker's discourse. Even the severe truths of religion, it was often conceded, could be enhanced by the speaker's expressive skill. Sometimes theatricality was deplored in the preacher and even in the man of politics, but there was a fairly widespread acceptance of the notion that communication—a necessary part of the cleric's, politician's, or advocate's activity—is improved by cultivation of some of the actor's skills. The need to win and retain an audience's attention was generally accepted, with the more serious debate being reserved for the question whether a speaker's efforts at artistry might lessen the moral force of his message. For obvious reasons, this question was felt to be more pressing in the case of clerics than for some other public speakers. But the non-Christian origin of most existing precepts on bodily communication, as well as Christian prejudices against the acting profession, should not be taken to indicate that *actio* in general necessarily aroused mistrust either among speakers in the pulpit or in their congregations. The ancient origins of the orator's eloquence, bodily as well as verbal, may not have been apparent to all listeners or observers, but for theoreticians these origins undoubtedly conferred dignity on the speaker's activity. In theories of painting, which I shall examine in the next chapter, a similar process was at work from the fifteenth century onward, and was turned to account in demonstrating the prestige which that art merited.

CHAPTER THREE

Acting and Visual Art

Diderot's *Lettre sur les sourds et muets*, as we have seen, discusses the *actio* of the stage performer in connection with the silent bodily communication of deaf mutes. But the vividness of the latter's gestures had much earlier been compared with the eloquence of characters in another silent art, painting, whose academic status the eighteenth-century actor often envied. Diderot's theory of acting was certainly influenced by his acquaintance with the visual arts, which began long before the start of his activity as art critic for the *Correspondance littéraire* in 1759.[1] He may well have encountered the analogy drawn by Roger de Piles in the commentary to his widely read translation of Dufresnoy's *De arte graphica* (1668), between the gestures of the dumb and the expressiveness of painting, before he wrote the *Lettre sur les sourds et muets*. His later *Essais sur la peinture* of 1766 offers some remarks on expression which seem indebted to de Piles, although Leonardo's *Trattato della pittura*, which had previously elaborated the same comparison, was probably also familiar to Diderot. The *Entretiens sur 'Le Fils naturel'* contains an urgent call for actors and playwrights to open themselves to the influence of the visual arts. Clearly, the mute art of *actio* bore certain similarities to the physical attitudes and expressions depicted on the painter's canvas or in the sculptor's marble, and there were many existing examples of actors being counselled to learn from such portrayals. The anonymous author of a *Lettre écrite à un ami sur les danseurs de corde et sur les pantomimes qui ont paru autrefois chez les Grecs et chez les Romains et à Paris en 1738* (1739) wrote, for instance, that actors in general should imitate the different bodily attitudes shown in painting and sculpture so that the spectator 'heard' the performer's silent action.[2] Nor was such advice directed only at players in the professional and secular theatre. In 1727 Franciscus Lang's *Dissertatio de actione scenica* had recommended that pupils in Jesuit colleges

[1] See Else Marie Bukdahl, *Diderot, critique d'art*, 2 vols. (Copenhagen, 1980 and 1982), I. 12, 30 (footnote 6).
[2] *Lettre écrite à un ami sur les danseurs de corde et sur les pantomimes qui ont paru autrefois chez les Grecs et chez les Romains et à Paris en 1738* (Paris, 1739), pp. 6-7.

(where there was a strong tradition of dramatic performance) should study the works of painters and sculptors as models for their own acting.

One of the most striking cases of visual art influencing the performance of serious drama, however, was unconnected with the actor's gesture and attitude. This concerned the costuming of characters on stage. Painters, whose academic training included the study of ancient history, had learnt to clothe their figures in historically appropriate garb many years before actors began to attempt a similar accuracy in the theatre. In his painting *Les Comédiens Français* Watteau depicts the performance of an apparently serious play in typically anonymous but recognizably antique environs, where the protagonists sport the anachronistic uniforms familiar in performances of neo-classical drama: panniers, bodices with lace trimmings, silk stockings, plumed hats, and the like. The exact subject of this painting has been a matter of much critical discussion, but most would agree that one of Watteau's intentions in the work is to mock the standard acting style of the Comédiens Français. Among the conventions which governed the latter's performance of serious drama, the correct reproduction of costume as it would have appeared in Phaedra's Greece or Nero's Rome found no place—until, that is, the early eighteenth century and the reforming efforts of Lekain and his fellows. But such correctness had long been seen as requisite in historical painting. What was there described as 'costume' did not, it is true, concern merely the depiction of clothing. According to Dézallier d'Argenville, writing in 1745, it comprised the exact rendering of mores, characters, fashions, customs, weapons, buildings, plants, and animals in the country the artist was attempting to depict (Locquin, pp. 165–6). But the correct reproduction of costume in the narrower sense of the word was often seen as paramount.

As the eighteenth century wore on, the scrupulous accuracy sought by the historical painter became a desideratum in the serious actor's performance too. Talma noted that Lekain had done all that was possible for his time in this regard (p. xvi). Talma himself, spurred by his acquaintance with artists, claimed to be a painter in his own right by virtue of his close attention to historical truth in costume (p. xviii). Earlier, Dorat wrote of the contribution made by Mlle Clairon to establishing the need for actors to wear clothes appropriate to their parts. 'Une Sarmate', Dorat observes, 'ne vient plus sur la scène faire l'amour en grand panier', nor do Roman heroes enter to declaim in periwigs and white gloves.[3] In 1790 Levacher de Charnois published his *Recherches sur les costumes et sur les théâtres de toutes les nations*, which according to the *Mercure de France*'s reviewer provided models for correct costume in the theatre which might equally be imitated by painters desirous of historical accuracy (December 1791, p. 88). But earlier in the century the painter's pre-eminence over the actor in this respect had rarely been questioned.

Of more interest to the present discussion are the many occasions when eighteenth-century writers counsel stage performers to study works of visual art for guidance in the striking of attitudes and the assumption of effective facial expression. In the *Encyclopédie*, for instance, the article 'Déclamation théâtrale' by Marmontel (later reproduced in his *Éléments de littérature*) provides specific examples of paintings and sculptures which may help the actor in fashioning his *actio*. Those who seek a model for passionate but noble gestures are advised to observe Reni's 'forces',[4] as well as the *Poetus*[5] and *Laocoön*. Marmontel also cites Timanthes's *Agamemnon*, Le Sueur's *St Bruno Praying*, Rembrandt's *Lazarus*, and Annibale Carracci's *Descent from the Cross* as examples of a 'jeu muet' for dramatists and actors to observe. Dubos had earlier referred actors to the ancient model of the *Dying Gladiator* (I. 360–1), and this work is similarly mentioned by the *Journal des théâtres, de la littérature et des arts* on 24 nivôse an VII (13 January 1798) (pp. 178–9). The actor and teacher Aristippe furnishes an extensive list of works the actor should study, but provides no specific examples of paintings to be imitated. He names Reni's figures for their touching expressions of pure, calm, and celestial love, Michelangelo's for their show of energy, pride, disdain, seriousness, stubbornness, and invincibility, and Rubens's for their fury, force, and excess.[6] However, some writers share Dorfeuille's

---

[3] Claude-Joseph Dorat, *La Déclamation théâtrale, poème didactique en trois chants, précédé d'un discours* (Paris, 1766), pp. 23–4. But with the Revolution, according to Étienne and Martainville, certain standards slipped: to please the ignorant sans-culottes, actors appeared in national colours when acting Greeks, Romans, Venetians, and Gauls, and 'Phèdre ne déclarait sa flamme à Hippolyte que la poitrine ornée d'une large cocarde tricolore' (Charles-Guillaume Étienne and Alphonse-Louis-Dieudonné Martainville, *Histoire du Théâtre Français depuis le commencement de la révolution jusqu'à la réunion générale*, 3 vols. (Paris, an X/1802), III. 142).

[4] i.e. presumably his four pictures of the *Labours of Hercules*, now in the Louvre but in Marmontel's day in the King's collection.

[5] i.e. the sculpted group *Poetus and Arria* in the Ludovisi collection in Rome, and of which Louis XIV commissioned a copy for Versailles.

[6] Aristippe [Bernier de Maligny], *Théorie de l'art du comédien, ou Manual théâtral* (Paris, 1826), pp. 330–2.

opinion that slavish imitation of a model in painting or sculpture is to be discouraged, on the grounds that the living quality of theatrical performance is incompatible with the fixedness of visual art. But one effect of advice such as Marmontel's is to emphasize the seriousness with which acting is viewed, as a profession in which the representation of masterpieces drawn from an art of liberal status is deemed appropriate as well as desirable. Besides, Talma was not the only actor to declare that a knowledge of the visual arts could with advantage be cultivated by stage performers—Mlle Clairon, for instance, writes that aspiring actors should take lessons in drawing (p. 270). Brizard and Bellecourt both served apprenticeships in the studio of Carle Vanloo before becoming actors, although in the preface to his *Salon* of 1765 Diderot writes that they, along with Lekain, had been *bad* painters who were forced by penury into the acting profession (*Salons*, II. 58). Previously to making his début with the fairground and boulevard entrepreneur Audinot the actor Corse had devoted himself to painting, and been taught by the neo-classicist Vien. One of the most famous Pierrots of the fairground, Antoine de la Place, had started painting in his youth, although little is known of his work. Finally, François Octavien, a contemporary of Watteau's, began his career in the elder Alard's fairground troupe. He painted military subjects and 'fêtes galantes', and was admitted to the Académie royale de peinture et de sculpture in 1725 with *La Foire de Bezons* (a subject, and title, possibly borrowed from Dancourt's play of 1695).

In a few cases there is even evidence of actors attempting to reproduce paintings on stage in the course of performing a play. With one celebrated exception, however, such imitations do not appear to have been carried out in a spirit of much seriousness. The exception is provided by a revival of Voltaire's tragedy *Brutus* in 1790. David's painting of the same name had been exhibited at the 1789 Salon, and itself drew on Voltaire's play (*see Plate 4*). In the revival the actor playing Brutus, delivering the last speech of the tragedy, adopted a pose unmistakeably resembling that of the main figure in David's painting, with the audience reportedly perceiving the allusion.[7] A less

---

[7] *Brutus* was first performed in 1730. See Robert L. Herbert, *David, Voltaire, 'Brutus' and the French Revolution: An Essay in Art and Politics* (London, 1972), pp. 77–8, and Gerhard Anton von Halem, *Paris en 1790, voyage de Halem* (Paris, 1896), p. 312. On David's involvement with the theatre see David L. Dowd, 'Art and the Theatre during the French Revolution: The Rôle of Louis David', *The Art Quarterly*, 23 (1960), 3–22. The anonymous author of the *Vérités agréables ou Le Salon vu en beau par l'auteur du Coup de patte* (Paris, 1789) observes: 'Cet ouvrage place M. David entre Shakespeare

clear-cut case of such imitation had earlier been provided in a production of Voltaire's tragedy *Les Lois de Minos* (first performed in 1774), which contained a scene where a temple door was broken down to reveal a tableau resembling Fragonard's painting *Corésus et Callirhoé* (described at length by Diderot in the *Salon* of 1765). Interestingly, Fragonard's work was itself based on a theatrical piece, either La Fosse's tragedy *Corésus et Callirhoé* (1703) or Roy's opera *Callirhoé* (1712).[8] Otherwise, precise pictural influences on *actio* in drama seem to have been of a more light-hearted kind than Diderot had envisaged when he wrote the *Entretiens sur 'Le Fils naturel'*. Three examples, all drawn from comedy, will illustrate the point. The earliest is provided by Carlo Bertinazzi's *Les Noces d'Arlequin*, which mimicked Greuze's *L'Accordée de village* in the same year as its exhibition at the Salon (1761). In a letter written on 8 November of that year to Count Durazzo, the Genoese diplomat who managed the court theatres in Vienna, Favart describes how at the end of the play—acted 'à l'impromptu'—the characters assembled to give a living representation of Greuze's picture, so exact that 'l'on croit voir le tableau même dont on a animé les personnages' (I. 200). The one alteration made to Greuze's composition was the added presence of Arlequin himself.[9] A later instance of the procedure occurs in Beaumarchais's *Le Mariage de Figaro* (1778). In the fourth scene of the second act three characters group to form a tableau which, as the stage directions tell us, exactly reproduces an engraving of Carle Vanloo's *La Conversation espagnole*, exhibited at the Salon of 1765. A final case is that of the one-act vaudeville by Jouy, Longchamp, and Dieulafoy, *Le Tableau des Sabines*, performed at the Opéra-Comique in 1800. At the end of this play, according to a note contained in the original text, the characters arranged themselves precisely after the model of David's painting *Les Sabines*, exhibited the previous year (*see Plate 5*). The fact that they have reproduced the painting, albeit parodically, is actually mentioned by one of the characters (Holmström, pp. 218-19).

et Corneille, et je ne sais de quoi je m'enorgueillirais davantage, ou d'avoir peint cette tête de Brutus, ou d'avoir créé le "qu'il mourût" des Horace. Il me semble que je retrouve également dans ces deux étonnantes pensées l'élan le plus vigoureux des âmes citoyennes et le sublime de la férocité' (p. 23).

[8] See Locquin, p. 236; also Prosper Dorbec, 'Les Premiers Contacts avec l'atelier du peintre dans la littérature moderne', *RHLF*, XXVIII (1921), 502.

[9] Greuze's pupil Mme Valori later wrote a one-act play, *Greuze, ou L'Accordée de village*, which was performed at the Théâtre du Vaudeville in 1813 (Bukdahl, I. 311).

In view of Diderot's strong words in the *Entretiens* about the desired proximity of painting to dramatic art, it is perhaps disappointing to record that no direct model in painting for his *drames* seems to exist. These plays were intended to depict the bourgeois world from which, as Diderot assumed, at least part of his audience would be drawn. (*Le Fils naturel* in fact received its première in an aristocratic private theatre.) Greuze exhibited his *Père de famille qui lit la Bible à ses enfants* in 1755, three years before the publication of *Le Père de famille*; but Greuze's picture, like its predecessors in genre painting by Jeaurat and Dutch seventeenth-century artists, depicted social types and situations distinctly more 'populaires' than those favoured by Diderot. The protagonists of *Le Fils naturel* show no signs of working for their living, and the middle-class 'père de famille' sanctions his son's marriage only when the intended daughter-in-law has been proven to be of equal social status.

It is true that the reverse influence, of drama on painting, was much in evidence in eighteenth-century France. So readily was painting linked with plays in the public mind that attending the Salons was sometimes itself described in terms of a visit to the theatre, such comparisons being particularly common towards the end of the century. In turn, the reactions of *amateurs* and connoisseurs to paintings were likened to the responses generally called forth in spectators by the various performing arts. Sometimes it was the 'faux connaisseur' whose appreciation was so characterized, which makes it clear that this attitude was regarded as inappropriate to the visual arts. They are not theatrical, it is suggested, and should not be treated as though they were. The *Journal de Paris* gives the following satirical description of one such visitor's reaction to the works on display in 1787:

Voyez-le agitant fortement sa lorgnette, il est là [au Salon] comme à l'Opéra, et il juge nos peintres avec les mêmes expressions, le même ton d'enthousiasme qu'il [*sic*] applaudissait autrefois aux productions de Rameau, qu'il dénigrait, il y a dix ans, la musique de Gluck, et qu'il la loue aujourd'hui. (*Journal de Paris*, 18 October 1787)

(Also present in this vignette, of course, is the notion that all of such a man's reactions are conditioned by fashion.) Other comments likening the Salon to a theatre are less precise, and are not necessarily disapproving. The *Observateur philosophique* of 1785 writes simply that 'plusieurs considèrent [le Salon] comme une espèce de spectacle', and implies that without this widespread view of its similarity to popular entertainment many would not trouble to pay it a visit (pp. 5–7). Importance is attached here to the accessibility of the visual arts to untrained intelligences; and, as we shall discover, this was also a contributory factor in the popularity of pantomime in eighteenth-century France. What is seen, it is again assumed, affects men more directly than what they perceive through their other senses. A much earlier report, in the *Observateur littéraire*, refers in a similar spirit to an analogy between the Salon and the theatre, and suggests in so doing that the immediacy of the effect which both institutions have on the beholder or audience is connected with their appeal to the sense of sight. According to this journal, the Salon resembles a theatre at the moment when the curtain is raised (vol. IV, 1759). Possibly as a response to the view that the theatre, like painting, speaks directly to men, some of the less serious commentaries on the Salons are themselves written in the dialogue form of plays, particularly of the popular vaudeville kind. In none of these cases, however, is anything made in the play itself of the links between drama and the visual arts.

Other criticisms describe paintings as though they were plays, and artists as actors, although they do not necessarily do so in order to emphasize the appeal of silent actions and images in the theatre. Moral scenes, in particular, are often called *drames*, and to judge by the following notice on Bilcoq's *L'Instruction villageoise* in the *Mémoires secrets* the *drame* continued to be popular throughout the 1780s: 'c'est encore de ces drames familiers faits pour attirer la foule; mais comme il est plus sagement traité que le premier [Wille's *La Mort du duc de Brunswick*], il ne saisit pas d'abord ainsi que lui de la curiosité générale; mais quand on en approche et que l'on le considère, on a peine à le quitter.'[10] In the *Observations d'une société d'amateurs sur les tableaux exposés cette année 1761*, which was printed in the *Observateur littéraire* that year, an analogy is drawn between artists and actors which implies the equivalence of their respective performances. According to the author, the painter Greuze enjoys the same kind of public favour as do those stage performers who offer audiences a 'plaisir familier'. Such artistes, it is suggested, inspire greater affection than actors who devote themselves to the elevated genre of historical drama, just as Greuze's everyday subjects exercise a more immediate appeal than the 'grandes machines' of the history painter.

[10] [Louis Petit de Bachaumont,] *Mémoires secrets pour servir à l'histoire de la république des lettres en France*, 36 vols. (London, 1777–89), XXXVI. 381–2. I am grateful to Mr Richard Wrigley for drawing my attention to this and other critical analogies between painting and drama.

Many writers on the Salons move beyond such general analogies between stage and canvas to suggest, usually damningly, that similarities exist between the bodily attitudes of actors and those of figures in a painting. Often the burden of criticism is the fact that an artist has reflected the affected posture of stage performers in his own attitudinizing characters. The false refinement of polite society which, in the opinion of some writers, has infected the movements of actors and dancers is stigmatized as a corrupting influence on painters. Diderot's *Salons*, which sometimes discuss the artist Carle Vanloo in terms of 'theatricality', are particularly harsh about the latter's depiction of lifeless gestures and attitudes which seem to have been borrowed from the stage or the dance-floor. In his dismissal of Vanloo's *Les Trois Grâces* Diderot remarks that 'Celle du milieu est raide. On dirait qu'elle a été arrangée par Marcel [a society dancing-master]' (*Salons*, II. 63).[11] (Diderot's allusion incidentally reveals that the interest in arts of physical movement which he expressed in writings like the *Lettre sur les sourds et muets* and the *Entretiens sur 'Le Fils naturel'* continued well into the 1760s.) In this Salon notice theatricality is identified with a lack of naturalism, of a kind which Diderot berates in the *Entretiens* when he discusses the stiffness of acting styles prevalent in the performance of serious French drama. A year later, in the treatise *De la poésie dramatique*, he contrasts the freedom of gesture in the Italians' acting style with that of the French, whose efforts are characterized by 'le raide, le pesant et l'empesé'.[12] In the review of Vanloo's painting it is the artificiality and manneredness of the beau monde which Diderot criticizes, as well as the anachronistic nature of the artist's depiction.

Another notice on this painting, contained in the *Journal encyclopédique* of November 1765, recalls that a commentator on the previous Salon (Diderot, though he is not mentioned by name) had attacked Vanloo's earlier version of the work, leading the artist to destroy it. The same commentator, it remarks, had advised Vanloo to show the Graces dancing 'une de nos contredanses ordinaires' (that is, not a society dance). The *Journal encyclopédique*'s critic mocks this suggestion —'Les connaisseurs prétendent que ce critique ne se connaît pas mieux en danse qu'en peinture'—and suggests that 'Il serait à désirer qu'on pût trouver sur nos théâtres le juste modèle des Grâces que

[11] See chapter 5.
[12] *De la poésie dramatique*, in Œ, p. 268. Subsequent references to *De la poésie dramatique* are to this edition.

M. Vanloo se proposait de peindre.' Exactly what is meant by this remark is uncertain. Is the writer referring to members of the Opéra, who were in theory meant to be actors and dancers as well as singers, or does he allude simply to performers in the non-lyric theatre? At all events, it is clear that a want of grace is being criticized. Diderot's objections to rococo manneredness are repeated much later in the century by other critics, and the deficiencies he perceives contrasted with the restraint, severity, and purity of line of the neo-classical style in painting. In 1787 criticisms of painting in Boucher's rococo idiom are still voiced in terms of its theatricality. The *Journal de Paris*'s review of the Salon that year notes that the successors to Poussin, Le Brun, and Lesueur, 'livrés la plupart à un genre de composition maniéré, ont eu l'air, pendant longtemps, d'en prendre les modèles moins dans la nature que sur nos théâtres, et ils ont transmis à la postérité, sans s'en douter, les afféteries de nos actrices.' Le Moyne and de Troyes, as well as Boucher, are blamed by this critic for the mischief, which was to be largely eradicated by the antique revival in the second half of the eighteenth century. This latter evolution, according to the same writer, was stimulated by the determined antiquarianism of the comte de Caylus.

The many comparisons which Diderot draws between the gestures and attitudes of actors and the depiction of human figures in painting often amount to a condemnation of bad acting. The same tendency is apparent in other critics too. One review of the 1777 Salon contains an unfavourable judgement on Lagrenée's painting of a scene from the life of Fabricius, in which the latter is shown refusing gifts sent him by Burrhus. The illustrious magistrate is dignified in his refusal, the critic writes, but his attitude is studied, and therefore utterly at odds with the true simplicity of this citizen–patriot. He looks like an actor trying to rise above his part.[13] Another commentator remarks of the same character that his attitude is unworthy of him, because 'Il est campé comme un personnage de théâtre', and has nothing of the Roman simplicity which existed in the days of good Fabricius.[14] Again, the rigidity of *actio* in non-comic theatrical performance seems to be at the origin of the criticism. A different note is struck in

---

[13] MS *Mercure de France, Exposition au Salon du Louvre des peintures, sculptures et autres ouvrages de MM. de l'Académie Royale* (see *Deloynes*, X, no. 191).

[14] *Lettres pittoresques à l'occasion des tableaux exposés au Salon en 1777*, letter IV (*Deloynes*, X, no. 190).

Lesuire's *Le Frondeur, ou Dialogues sur le Salon* of 1785, where Berthélemy's *Torquatus* is found reminiscent of a histrion: 'Tourmenter ses yeux, sa bouche et ses bras, comme fait un mauvais acteur durant une ritournelle, ce serait blesser la vérité; cacher la moitié de sa figure et lui prêter un geste emporté, c'est manquer de noblesse et dénaturer le sujet' (*Deloynes*, XIV, no. 329). The depiction of an over-emphatic attitude on canvas, in other words, is as reprehensible as wild gesticulation in the theatre. Elsewhere critics liken a painter's picture to a scene acted out by buffoons, or 'acteurs d'un théâtre de campagne'. In 1779 Lagrenée's offering at the Salon is found by one critic to be unsophisticated in its portrayal of characters. He judges the painting itself to be adequately composed, but says that it is wanting in expression and colour, and that 'tous ces personnages manquent de caractère; ils ressemblent à des comédiens bourgeois.'[15] And Suvée's *Coligny*, exhibited at the Salon of 1787, is considered caricatural by one commentator, who says that Coligny 'a l'air et le costume d'un Gilles de théâtre.'[16] A review published by the *Correspondance littéraire* in October 1779 finds the great historical figures portrayed by Lagrenée the elder, Duramaau, Lépicié, Brenet, and Renou inadequate because they display rustic crudeness rather than the elevation thought proper to them. What is there to say of all these works? That M. Lagrenée's Popilius is no Roman ambassador, that the Antiochus whom this miserable ambassador has trapped inside a magic circle is no king, that the characters look like country actors.[17] A similar criticism had been voiced several years earlier by Diderot with respect to the portrayal of Hector in Challe's painting of Helen and Paris (*Salons*, II. 85).

Even when a painting or sculpture has been skilfully executed, critical comparison of its figures with performing artistes is usually tantamount to a condemnation of the work. This is notably true when the comparison is with dancers, who are generally described as mannered and intent on playing to the audience. The *Correspondance littéraire*'s review of the 1779 Salon, considering Clodion's sculpted figure of Montesquieu, asks rhetorically whether the artist's model was not a dancer from the Opéra, with his precious, affected pose. 'Et c'est là le portrait du sage qui a rendu à l'humanité ses titres

---

[15] *Ah! Ah! Encore une critique du Salon! Voyons ce qu'elle chante* (*Deloynes*, XI, no. 208).
[16] *Tarare au Salon de peinture* (*Deloynes*, XV, no. 376).
[17] Grimm, Diderot, Raynal, Meister, etc., *Correspondance littéraire, philosophique et critique*, edited by Maurice Tourneux, 16 vols. (Paris, 1877–82), XII. 323.

qu'elle avait perdus!' (XII. 309.) Stronger still is the criticism of paintings which recall vulgar types of popular entertainment, without the approximation to nobility that theoretically accompanied performance at the Opéra or the Comédie-Française. A notice on the 1787 Salon remarks of a character in Wille fils's *La Mort du duc Léopold de Brunswick* that 'Le prince a l'air d'un mannequin, d'un sauteur de chez Nicolet', whereas according to the critic 'Ce projet patriotique [était] si propre à enflammer le génie.'[18] The acrobats and tumblers of Nicolet's playhouse are dismissed as unsuitable models for serious historical painting; and so, presumably, they would have been for the acting of *drames* as Diderot envisaged the genre in the *Entretiens sur 'Le Fils naturel'*. Diderot's scathing notice on Francisque Millet's *Paysage où Sainte Geneviève reçoit la bénédiction de Saint Germain* in the Salon of 1765 includes the observation that it resembles a scene from comic opera (*Salons*, II. 118), and of Roslin's *Un Père arrivant à sa terre, où il est reçu par sa famille* he confesses that

> Une idée folle dont il est impossible de se défendre au premier aspect de ce tableau, c'est qu'on voit le théâtre de Nicolet, et la plus belle parade qui s'y soit jouée. On se dit à soi-même: Voilà le père Cassandre; c'est lui, je le reconnais à son air long, sec, triste, enfumé et maussade. Cette grande créature qui s'avance en satin blanc, c'est Mlle Zirzabelle, et celui-là qui tire sa révérence, c'est le beau M. Liandre. (Ibid., p. 124)[19]

The comparison is intended to imply a crudeness and indelicacy of characterization in the picture which resembles the *actio* of Nicolet's fairground and boulevard performers. It signifies a breach of the decorum that was meant to govern history painting. Many reviews of the Salons draw attention to the way in which a theatrical quality degrades work that is intended to be elevated and noble. 'Theatrical' art is seen to lack subtlety, and the 'actors' to be deficient in grace or a sense of propriety.

Whether this view of theatricality is mere prejudice, the residue of many years' disrespect for the acting profession, or whether it still reflects the overall quality of acting style, it is impossible to state with certainty; but the available material suggests that tradition played a large part in the forming of such adverse opinions. Diderot's own

---

[18] *Tarare au Salon de peinture*, part II (*Deloynes*, XV, no. 377).

[19] On Nicolet, see also chapter 4. The painter Pierre criticized David's *Brutus* for the fact that three figures were placed on a line, while the principal actor was in the shadow: he likened the effect to the kind produced at Nicolet's theatre (Jules David, *Le Peintre Louis David* (Paris, 1880), p. 37).

respect for the actor was considerable, as works like the *Entretiens sur 'Le Fils naturel'* and the *Paradoxe sur le comédien* make clear, but condemnation of paintings because of their theatrical quality is commoner in his Salon criticism than in that of any other eighteenth-century writer. On one occasion, it is true, he praises a painting because it depicts a character behaving as an actor would, but what he appreciates here is simply the psychological accuracy of the portrayal. Commenting admiringly on Hallé's *Course d'Hippomène et d'Atalante*, he remarks that

La victoire ne peut plus lui [Hippomène] échapper; il ne se donne pas la peine de courir; il s'étale, il se pavane, il se félicite: c'est comme nos acteurs, lorsqu'ils ont exécuté quelque danse violente; ils s'amusent encore à faire quelques pas négligés au bord de la coulisse. C'est comme s'ils disaient aux spectateurs: Je ne suis point las; s'il faut recommencer, me voilà prêt: vous croyez que j'ai beaucoup fatigué, il n'en est rien. Cette espèce d'ostentation est très naturelle, et je ne souffre point à la supposer à l'Hippomène de Hallé. (*Salons*, II. 85)

On this occasion, in any case, there is some critical dissent from his opinion. Grimm objects that the theatrical model was an inappropriate one for Hallé to have chosen: for a Greek raised amidst woods and mountains to strut about like Vestris or Gardel (celebrated dancers of the time) at the Opéra is false and in bad taste (ibid.). But Grimm's interpolation does confirm the view shared by many commentators on the performing arts in the eighteenth century, namely that where *actio* was mannered or contrived it was justly criticized.

The critic of the *Année littéraire* wrote against a picture exhibited in 1783 by Ménageot that the attitudes of its characters, which were stiff and unnatural, seemed to have been dictated by theatrical rather than painterly considerations. Like Dorfeuille, he went on to argue that the conventions of the two art-forms were in many respects dissimilar, and that a dramatic model might be an inappropriate one for the artist to observe.[20] But the same criticism as he levelled at this work might also be voiced with reference to David's much more famous painting of 1784, *Le Serment des Horaces*. Its inflexibly linear structure was indeed found objectionable by some contemporary commentators.[21] The work is of interest here in connection with the theatri-

[20] See chapter 6.
[21] See, for example, the *Supplément du peintre anglais*, quoted in Thomas Crow, 'The *Oath of the Horatii*: Painting and Pre-Revolutionary Radicalism in France', *Art History*, I (1978), 433; also Seymour Howard, *Sacrifice of the Hero: The Roman Years. A Classical Frieze by Jacques-Louis David* (Sacramento, 1975), p. 85.

cality of eighteenth-century French painting, as are his *Brutus* and *La Mort de Socrate*, in that they appear to meet a requirement voiced by another Salon critic, who suggests in 1787 that historical painters should draw their subjects from ancient and modern drama because 'les plus beaux sujets d'Homère, de Virgile et du Tasse sont épuisés. Les belles situations théâtrales seraient certainement très propres à produire de grands effets en peinture.'[22] Young artists at the École des élèves protégés, which was established in 1749 under the general control of the Académie royale de peinture et de sculpture in order to nurture the talents of especially gifted pupils,[23] were set to compose pictures based on scenes from plays, such as *La Mort d'Hippolyte* after Théramène's récit in Racine's *Phèdre*, or *Les Grâces enchaînées par l'Amour* after a scene in Sainte-Foy's play *Les Grâces* (Locquin, p. 88). The origins of David's *Le Serment des Horaces* are disputed. Corneille's *Horace* was certainly not his only source, but it seems equally clear that it was one of his inspirations.[24] Significantly, a ballet by Noverre based on the same story has also been mooted, although it seems inherently unlikely that the rigid pose of David's characters is the reflection of a similar stiffness in the dancers of *Les Horaces et les Curiaces*. The debt of David's *Brutus* to Voltaire's tragedy has already been mentioned. Finally, his 1787 painting of *Socrate au moment de boire la ciguë* is in a tradition on which more than one eighteenth-century dramatist drew, but there is no firm evidence that David borrowed anything specifically from the performance of any of their plays.[25]

Italian comedy, with its rich tradition of gestural acting, undoubtedly influenced some seventeenth- and eighteenth-century French artists. The work of Watteau and his master Gillot comes to mind in this respect. Various literary sources have been suggested for the former's *Retour de l'île de Cythère*,[26] although recent evidence indicates

---

[22] *Observations critiques sur les tableaux du Salon de l'année 1787* (Paris, 1787), p. 17.
[23] See Louis Courajod, *L'École royale des élèves protégés* (Paris, 1874).
[24] See Anita Brookner, *Jacques-Louis David* (London, 1980), p. 69 ff.
[25] See Jean Seznec, *Essais sur Diderot et l'antiquité* (Oxford, 1957), pp. 19–20. Various influences shaped David's composition. The Oratorian père Adry advised him on the most effective manner of conveying Plato's grief, and suggested that he consult an ancient depiction of the death of Meleager for a model of the appropriate bodily attitude. See E. Bonnardet, 'Un Oratorien et un grand peintre', *Gazette des beaux-arts*, I (1938), 311–15. An early nineteenth-century witness reports that 'Le père Lacouture, à la barbe longue et touffue, servit de modèle à notre peintre David pour sa mort de Socrate' (*Mémoires de Ch. de Pougens* (Paris, 1834), p. 27).
[26] See Robert Tomlinson, *La Fête galante: Watteau et Marivaux* (Geneva, 1981), pp. 111–13.

that a sketch of Gillot's, possibly of a scene from Dancourt's *Les Trois Cousines*, may have been a principal inspiration.[27] Watteau's painting of *Arlequin empereur de la lune*, after Nolant de Fatouville's three-act comedy (which Watteau probably saw performed at the foire Saint-Laurent in 1707), is one of his very few canvases to reproduce a scene from drama faithfully. The lost painting now known from an engraving by Charles-Nicolas Cochin, *Pour garder l'honneur d'une belle*, is of the last scene of a comedy by Pierre-François Biancoletti, *Arlequin fille malgré lui* (performed at the foire Saint-Laurent in 1713). These two works alone refute Diderot's remark in the *Essais sur la peinture* that 'On n'a point encore fait, et l'on ne fera jamais un morceau de peinture supportable d'après une scène théâtrale.'[28] Diderot meant that observation as a cruel indictment of acting, theatrical décor, and even dramatists. But it is much commoner in Watteau to find an amalgamation of the seen and the imagined than to encounter direct imitations of this kind. The picture of *Pierrot content*, for instance, does not appear to be an illustration drawn directly from a play, although the character clearly belongs to Italian comedy. There is a similar uncertainty about the origin of his *Les Comédiens Français*. According to the *Mercure de France* of December 1731, this picture shows French actors performing a tragi-comedy. Later commentators have disagreed about the type of play that is being acted: for some it is a tragedy (possibly *Andromaque* or *Bérénice*) and for others a comedy, while for others again the painting is an allegory of the French theatre. The picture of *Les Comédiens Italiens*, equally, seems not to show an image drawn from live theatre. At most, the posed attitudes of the characters suggest the beginning or end of a play rather than a moment in performance. The *Mercure de France* declared in 1733 that the work was composed of portraits of people 'skilled in their art', whom Watteau had simply clothed in the costumes of the different *Commedia dell'arte* actors. Once more, Watteau apparently constructed a theatrical canvas from scattered studies of friends, and costumed them at will. Five preparatory sketches for this painting exist, and suggest that the artist composed his work piecemeal. His fondness for depicting theatrical scenes, then, arose from a desire not to describe, but to evoke: the theatre was for him a means rather than an end. The comte de Caylus disapproved of Watteau's random compositional

---

[27] I am indebted to Mme Marianne Roland Michel for this information.
[28] *Essais sur la peinture*, in Œ, p. 713. Subsequent references to the *Essais sur la peinture* are to this edition.

method, but was mistaken to claim, as he did in his life of the painter, that Watteau never produced complete studies on canvas.

Although Diderot's dramatic theory has affinities with the practice of the *Commedia*, his *drames* called for greater solemnity in performance than the Italians were accustomed to show. The spirit of the *drames* was, however, admirably captured in the work of another painter, Greuze. In *L'Accordée de village*, the sensation of the 1761 Salon, the characters enact the special ritual of speaking gesture and solemn word which Diderot had surely envisaged for the performance of *Le Fils naturel*. The play was not publicly staged until 1771, but the stir created by the appearance of the text together with the *Entretiens* in 1757 could have influenced Greuze in his choice of subject. (At the same time, it must be owned that Greuze's work is distinctly more erotic than Diderot's *drames*.) *Le Père de famille*, first performed in 1761, may similarly have provided a model for a later painting, Aubry's *L'Amour paternel* (exhibited in 1765). Finally, whether or not David drew anything for his Socrates painting from Diderot's plan for a *drame* on the death of the philosopher, contained in *De la poésie dramatique* (pp. 198–9), Diderot himself certainly thought his own sketch had strong pictorial possibilities. He envisaged the projected scene of Socrates' last moments as 'une suite de tableaux, qui prouveront plus en faveur de la pantomime que tout ce que je pourrais ajouter' (p. 272).

One of Diderot's main preoccupations in developing the theory of the *drame* was the possibility of conveying reality, such as traditional tragedy and comedy seemed to prohibit, on the stage. This desire, too, was in theory to be realized in the *actio* of the performers as well as in their verbal utterances. Here, it may be felt, comparisons with the artist's portrayal of characters on the canvas were bound to break down. In painting which deals with actions (the province of historical painting in the eighteenth century), and which must choose the 'fruitful moment', bearing the greatest significance, for depiction on canvas, the temptation for the artist to exaggerate in order to be assured that his 'meaning' will be unequivocally conveyed is obvious. In drama, which moves in time, and normally has words with which to reinforce figural depictions, that temptation seems more readily avoidable. Diderot's own difficulties as theorist of drama are obvious. On the one hand he wanted fewer words and more *actio* in the theatre. On the other he surely realized that since not all gestures are themselves charged with evident meaning, they often require verbal elucidation. Additionally, in *Le Fils naturel* Diderot tried to create a form of drama that avoided 'theatrical' enlargement by confining itself to the small world of everyday life. His play, he notes in the *Entretiens*, should be judged as a private spectacle, unfolding 'dans le salon de Clairville', and having no need of the kinds of exaggeration made necessary by the size of conventional theatres. It is a domestic commemoration of virtuous deeds, and should be performed in accordance with that fact. But much in the play strikes the modern reader, as it struck Diderot's contemporaries, as larger than life and grandiloquent.

The desire for naturalness allied with didacticism lies behind one of the proposals made for the new kind of drama in the *Entretiens*, that it should contain tableaux. A principal function of the tableau as conceived by Diderot and those he influenced was to underline the peaceful ordinariness of stage action. It is contrasted by Dorval in the *Entretiens* with the abruptness and 'staginess' of the coup de théâtre: 'J'aime bien mieux les tableaux sur la scène où il y en a si peu, et où ils produiraient un effet si agréable et si sûr, que ces coups de théâtre qu'on amène d'une manière si forcée, et qui sont fondés sur tant de suppositions singulières' (p. 88). The tableau arises from the preceding action, emphasizes continuity rather than interruption, and provides a moment of repose in which the significance of what has gone before can be underlined. Furthermore, it has the effect of drawing together the different strands of dramatic action, uniting them in a harmonious whole which observes the same laws of compositional unity as the beholder seeks in a well-made picture.[29] It thus has both a moral and an aesthetic function. Diderot's tableaux emphasize the rhetorical principle of *compositio* on a small scale, and in so doing call

[29] The interesting notion of 'papillotage', which Marian Hobson develops in *The Object of Art* (Cambridge, 1982), is perhaps related to this aesthetic. 'Papillotage', on this interpretation, is replaced in eighteenth-century painting by the straightforward depiction of empirical reality. It might be added that the various writings of Diderot, Rousseau, Mercier, and Noverre (among others) reveal the extent to which the disunity evident in some art-forms—especially the performing ones—was disliked during this period. For Diderot at least, the argument seems to be concerned, on the one hand with reality, and on the other with certain kinds of artistic virtuosity in depiction which are unreliable guides to the real and true. As such it has affinities with Plato's distinction between sense and intellect: Plato holds that the senses are to be construed as dealing with the changeable, and that reality is unchangeable, so that the senses cannot give men knowledge of reality. On the tableau in visual art and drama, see also Michael Fried, *Absorption and Theatricality: Painting and Beholder in the Age of Diderot* (Berkeley, Los Angeles, and London, 1980), chapter 2, especially p. 76 ff.

on the actors' ability to convey through bodily attitude the same kind of integration into a greater whole as Diderot demands of the parts of a painting (*Œ*, p. 790). *Le Fils naturel* and *Le Père de famille* are concerned with the virtues of domestic unity, and their tableaux allow the moral moments of fraternity and communion to be underlined.

Both Beaumarchais and Mercier were strongly influenced by Diderot's theory of *drame*, and in Mercier's theoretical writings the moral and aesthetic harmonies achieved through *compositio* are emphasized. The tract *Du théâtre* insistently states the need for the playwright to present an ensemble, rather than individuals, in his *drames*, and this requirement is described in painterly terms:

> Il ne s'agit point dans la comédie [probably in the general sense of 'drama'] de faire des portraits, mais des tableaux. Ce n'est pas tant l'individu qu'il faut s'attacher à peindre, que l'espèce. Il faut dessiner plusieurs figures, les grouper, les mettre en mouvement, leur donner à toutes également la parole et la vie. Une figure trop détachée paraîtra bientôt isolée: ce n'est point une statue sur un piédestal que je demande, c'est un tableau à divers personnages.[30]

Diderot had earlier declared that the time was ripe for dramatists to depict groups rather than individuals, representatives of social types rather than exceptions to the general rule. Mercier states his beliefs still more firmly: he wants the poet to show him the world's stage, not the sanctuary of a single individual. Anyone who has reflected on the mind, behaviour, and character of the different men he has met will show them interacting, not in isolation. What gives life to drama and lends weight to its moral teaching is the simultaneous and reciprocal action of all the characters (ibid.). Mercier's views on the art of characterization rest on narrative principles. A gradual rather than an instantaneous depiction is required, and the emphasis is on the playwright's providing cumulative details that reveal the nature of an individual. (The reader may be reminded of Diderot's praise in the *Éloge de Richardson* for the English writer's skill in this respect, and of his likening Richardson's novels to *drames*.) Mercier also stresses the need for an integrated portrayal of character in the dramatist's work, and again expresses his opinion in the language of painting: 'J'insisterai toujours à représenter que les caractères des hommes sont mixtes, qu'un ridicule ne va jamais seul, qu'un vice ordinairement

---

[30] L.-S. Mercier, *Du théâtre ou Nouvel Essai sur l'art dramatique* (Amsterdam, 1773), p. 69.

est étayé par d'autres vices, que vouloir détacher un défaut de ceux qui l'environnent et l'avoisinent, c'est peindre sans observer la dégradation des ombres et des couleurs' (p. 71). Finally, Mercier's encouragement to his fellow-playwrights to portray multitudes, not minutely, but with broad movements, recalls the art critic's reference to painters (like Chardin, as Diderot describes him in the *Salons*) whose technique is one of free brushstrokes, whose 'faire' is 'large':

saisissez les grands traits, vous aurez de larges coups de pinceau à donner, vous recontrerez des caractères expressifs et variés; vous vous servirez malgré vous de teintes vigoureuses; jamais vous n'éprouverez cette stérilité qui gagne le bel esprit . . . Qu'on vous appelle peintres à la grosse brosse, qu'importe; on a fait le même reproche à Molière; on l'a blâmé par ce qu'il a aujourd'hui de plus précieux. Poursuivez vos tableaux, et laissez vos rivaux fatiguer leur vue à faire des miniatures de poche. (pp. 80–1)

Diderot's objection to the coup de théâtre in drama was based on its theatricality, or over-insistence; but a possible objection to the tableau is that it does not insist enough. To define the quality of 'the dramatic' is not easy, but it seems to depend on a sense of urgency which is not a necessary part of pictorial art. This is not to say that urgency has to be continuously conveyed in drama. It is a part of the composition, emphasized at some points in the action and put in abeyance at others. Action is crucial to plays, but not to paintings, whose function may be (and in the eighteenth century often was) purely decorative. This fact suggests a practical danger in the playwright's and actor's observation of models from visual art. Diderot conceived of the tableau as being not merely picturesque, but also persuasive, a tool with which to make a moral statement. But a potential drawback of the tableau was that, in slowing down or halting action in a performance, it might work against the overall effort at persuasion. Diderot recognized this danger, and tried to avert it by confining such effects to the beginning and end of acts in a play. The theory beind this expedient is developed in a letter to Mme Riccoboni, who had insisted that the theatre should be a 'tableau mouvant', and found his ideas about drama misguided on those grounds.[31] Not all playwrights heeded the warning, however. In 1772 Cailhava unwittingly gave an example of the weakness of some pictorial effects in drama in his discussion of Sainte-Foy's short comedy *Les Grâces*. He

---

[31] See *Œuvres complètes de Diderot* (subsequently *A.-T.*), ed. Jules Assézat and Maurice Tourneux, 20 vols. (Paris, 1875–7), VII. 396–9.

adjudged one tableau in the play worthy of being copied by painters:[32] it showed a pastoral scene in the style of Boucher, Cupid chained to the foot of a tree with garlands of flowers, and nymphs seated around him. Yet Cailhava's reader is struck by the undramatic, purely decorative quality of the tableau, which closely resembles the rococo painting by Carle Vanloo severely criticized by Diderot in the *Salon* of 1763, and subsequently destroyed by the artist. Being a motionless scene, it offers little opportunity for the gestural acting of which Cailhava has just signalled his approval. The resources of drama, in other words, are not necessarily commensurate with those of the (more or less static) visual arts.

But, in conclusion, it should be emphasized that some of the most effective paintings based on dramatic performances in the eighteenth century appear to the innocent eye to be non-theatrical, in the strict sense that it is hard to imagine the artist's depiction as being of a scene which was actually enacted before a theatre audience. Carle Vanloo's celebrated portrait of Mlle Clairon as Médée (a painting disliked by Diderot for the 'theatrical' falseness of its décor) is a good example of this phenomenon (*see Plate 6*). This is not to say that it is innocent of details that fit it into a dramatic tradition, although it suggests the theatrical setting less strongly than does a painting by Charles-Antoine Coypel of *Médée et Jason* (1715) which seems to have been a model for Vanloo. But, in general, it illustrates the way artists might develop and embellish, according to the dictates of their own art, an image momentarily present in performance. With rare exceptions, critics found Médée's facial expression a fittingly passionate one, and preparatory sketches for the work show it to have been well within the conventions of 'têtes d'expression' followed by both painters and actors of the period. An account contained in the *Observations sur l'exposition de peintures, sculptures et gravures du Salon du Louvre* which was published in the *Observateur littéraire* for 1759 suggests that it struck beholders as an expression drawn from the theatre: 'La tête de Médée est le portrait non pas de la personne seulement de Mlle Clairon, mais de Mlle Clairon actrice, excitant encore sur la toile une partie des passions qu'elle agite fortement sur la scène.'[33] But despite such testimony, the painting does not convince as a representation of

---

[32] Cailhava de l'Estendoux, *De l'art de la comédie*, 4 vols. (Paris, 1772), I. 421.
[33] *Deloynes*, XLVII, no. 1259: *Observations sur l'exposition des peintures, sculptures et gravures du Salon du Louvre, tirées de l'Observateur littéraire.*

properly dramatic action. Some writers of the time were struck by its painterly quality of immobility, contrasted with the movement that characterizes dramatic performance. Médée and Jason are found awkward and 'posed' by more than one critic. For the author of the *Lettre critique à un ami sur les ouvrages de MM. de l'Académie exposés au Salon du Louvre 1759* (*Deloynes*, VII, no. 90), Médée looks constrained. Neither her face nor her attitude betrays the emotion which such appalling vengeance should provoke, but, rather, both make her appear stiff and embarrassed. Another writer, the anonymous author of the *Lettre d'un artiste sur le tableau de Mlle Clairon*, considers that only her arms are alive, and even their action is constricted by the settledness of her pose. Her head and shoulders, he writes, are like marble, and so straight as to destroy any sense of motion.[34] On the other hand, he continues, this immobility may be justified by the argument that Médée is now at rest. Her fury has been assuaged, and she can contemplate her success (p. 944). Such an interpretation certainly seems applicable to the preparatory sketch (now in the Musée des Beaux-Arts at Pau: *see Plate 7*), which shows a very different Médée from that of the finished version: in the former she indeed radiates calm and tranquillity, but the same is not true of the completed picture at Potsdam. Jason's rigidity is criticized too by some observers. The author of the *Lettre critique* remarks that he seems not to be driven by the impetuous rage which the subject calls for, but appears fearful and 'n'avanc[e] qu'à pas mesuré' (p. 7). The anonymous manuscript *Portrait de Mlle Clairon, par Carle Vanloo* observes, however, that this can readily be explained: Médée has cast a spell on her husband to render him motionless.

Finally, it was reported by one witness that Mlle Clairon herself declared Vanloo to have surpassed the art of the theatre in his painting, and that she 'ajout[a] poliment *qu'il faudra que l'original étudie longtemps sa copie*' (*Lettre d'un artiste*, p. 948). The preliminary studies show Vanloo's efforts to find the best *pictorial* solution to the problems of presenting his subject, and there is no direct evidence that the final work mimetically reproduced a moment from Mlle Clairon's performance of the part in Longepierre's play. On the other hand, the last version appears in comparison with the Pau sketch eminently theatrical in a broad sense of the term, that is, in its depiction of emphatic facial expression and gesture. Diderot's suggestions for recasting the figure of Jason are themselves highly melodramatic.

[34] *Lettre d'un artiste sur le tableau de Mlle Clairon* (Paris, 1759), in *Deloynes*, VII, no. 90.

Acting and Visual Art

Finding Jason insufficiently impassioned, he writes that 'Il fallait lever au ciel des bras désespérés; avoir la tête renversée en arrière; les cheveux hérissés; une bouche ouverte qui poussât de longs cris, des yeux égarés' (*Salons*, I. 64). As the above discussion has tried to show, the nuances by which the word 'théâtral' and its cognates may be coloured are many. In some cases their use is pejorative, in others not. The 'classical' dignity of David's *Le Serment des Horaces*, to take a different example, makes its observer unwilling to apply so potentially debasing a description to it, but few would deny that the work is a remarkably 'theatrical' one, whatever the precise sources on which David drew for his composition.

It seems to have been the view of eighteenth-century critics that paintings may be weakened if influenced by certain forms of dramatic performance, especially that part which relates to action and attitude. About the reverse influence it is difficult to be precise, largely because contemporaries rarely offer a reasoned aesthetic judgement of the appropriateness of pictorialism in the enacting of a play. The evidence suggests that the occurrence of tableaux was felt by some, like Mme Riccoboni, to be potentially damaging in drama, and by others to be a welcome development. The tableau, of course, represents only one form of visual effect in theatrical performance: the following two chapters examine some others in relation to *actio*.

CHAPTER FOUR

Pantomime Performances

An anonymous and unpublished play performed in 1799 at the Théâtre Sans Prétention sums up its author's view of the excessive devotion to pantomime which some of his contemporaries displayed.[1] Mondor, whose daughter Sophie is of marriageable age, has decided to promise her to the author of the most successful work in this genre submitted to him, for, unlike Sophie, he believes that such success will be a mark of genius. Sophie holds that actors rather than playwrights are the real creators of pantomime. If an actor were cold, she observes, yet his role called for passion, the sense of his action would be impenetrable to an audience, and in that case the author's instructions would be of no avail. In the end Mondor concedes that the art of the actor adds to the effectiveness of a well-made plot, for 'Quand un acteur est costumé suivant le rôle qu'il représente, qu'il a du feu et les gestes vrais, je me figure dans mon esprit les plus belles paroles du monde, en vers, en prose, comme je veux.' A similar observation had been made by another author, Mercier, shortly before the performance of *La Pantomimanie*, and I discussed its implications in chapter 1. Not everyone agreed that the art of wordless communication in the theatre was a supreme form of dramatic art, however. Sophie's maid Lisette speaks for the unbelievers: 'Si dans ces pièces on parlait encore, on pourrait voir l'esprit de celui qui l'a faite, mais ne faire que des gestes et des mines, on a l'air fou. Que ne vous fait-il épouser une machine, elle ne ferait du moins que les mouvements que vous voudriez.' And her mistress remarks that the composition of this type of dramatic work is usually a sign that its author lacks the talent to do greater things. Although she would not refuse to marry a man simply on the grounds that he had written a pantomime, nevertheless 'il n'y a que ceux qui ne se sentent ni assez d'esprit ni assez de style pour faire parler convenablement leurs personnage qui s'amusent à faire ces pièces, ou des jeunes gens pour attraper quelque argent.'

[1] *La Pantomimanie*, contained in the *Répertoire du Théâtre sans Prétention*, 1798–1805 (held by the Bibliothèque historique de la Ville de Paris, 611133).

These exchanges, and others in the play, reveal some important aspects of the debate about acting in eighteenth-century France: one, that the fondness for pantomime was indeed regarded as a mania by many contemporaries; another, that the emphasis placed on non-verbal communication in the theatre was not universally welcomed; and a third, that a taste for pantomime was apparent in the educated and well-to-do (Mondor is of solid bourgeois stock) as well as among the 'petit peuple' to whom wordless entertainment was obviously attractive.

The entertainment offered at the fairs and in the boulevard theatres of eighteenth-century Paris presented several striking examples of spectacles more or less closely related to pantomime. According to de La Porte's almanac *Les Spectacles de Paris* of 1754, plays were not performed at the fairs until 1678. The first one on record, *Les Forces de l'amour et de la magie*, was a 'mélange assez bizarre de sauts, de récits, de machines et de danses. Ces sortes de pièces étaient représentées par des sauteurs qui formaient différentes troupes.'[2] Such entertainments, and their later development by entrepreneurs like Nicolet and Audinot, were complemented by a variety of non-performing spectacles. The almanacs of the day record the exhibition at fairs, for the crowd's amusement, of curiosities like dwarves and giants, menageries of animals, and 'cabinets' containing automata and devices for creating different kinds of optical effect. Similar offerings were available in the boulevards (starting with the boulevard du Temple, and subsequently extending to the boulevard Saint-Martin), which gradually supplanted the fairs in public favour. The number of boulevard theatres built, and the variety of entertainments offered there, greatly increased after the Revolution and the edict proclaiming the 'liberté des théâtres'. Before 1789 only five theatres existed on the boulevard du Temple: Nicolet's Grands Danseurs du Roi (which later became the Théâtre de la Gaîté), Audinot's Ambigu-Comique, Sallé's Théâtre des Associés, the Théâtre des Délassements Comiques, and the Théâtre du Lycée Dramatique, built for the use of pupils training for the Opéra. When restrictions were lifted, theatres burgeoned on the boulevards.

---

[2] [Abbé Joseph de La Porte,] *Les Spectacles de Paris, ou Suite du calendrier historique et chronologique des théâtres* (Paris, 1754), p. 166. Maurice and Alard were the creators and directors of the troupe which performed this work: see Arnould-Mussot's anonymous *Almanach forain, ou Les Différents Spectacles des boulevards et des foires* (Paris, 1773), unnumbered pages.

Outside the fairs and boulevards, there were other contributions to the picture of popular entertainment which had connections with pantomime The pleasure-dome of the Colisée, a vast complex of amphitheatre, surrounding galleries, and gardens on the Champs-Elysées, offered many of the amusements associated with the small theatres, such as pantomime proper and the fairground art of rope-dancing.[3] Gachet's *Observations sur les spectacles en général, et en particulier sur le Colisée* of 1772, though excited by the possibilities which the development offers, none the less contains a tepid appreciation of the way the art of pantomime has been treated there. The first performance of a pantomime called *La Fête chinoise*, 'par le lointain, et de plus par le mauvais jeu des acteurs, le défaut des machines, la ridicule imitation de la foudre, n'a pas satisfait . . . l'attente des spectateurs'.[4] Among the pantomimes mounted at the Colisée, with accompanying fireworks, were *Don Quichotte* (October 1771), *L'Entrée de l'ambassadeur de la Chine* (July 1772), and *Le Temple de mémoire* (September 1772). According to various contemporary commentators, the Colisée was intended to be a centre for all classes to visit and enjoy. Certainly it attracted the highest in the land, and was visited successively by Marie-Antoinette, the princes of France, and the (incognito) Emperor of Austria.[5] A second area created for the pleasure of Parisians, both well-born and humble, was the summer Vaux-Hall (which had various imitators), where Torré's spectacular firework displays, or 'pyripantomimes', were staged.[6]

Another practitioner of quasi-pantomimic art outside the fairs and boulevards was the stage designer Jean-Nicolas Servandoni, possibly the uncle of Servandoni d'Hannetaire.[7] Near the end of the eighteenth century Lablée described Servandoni's type of pantomime as ideally suited to the republican fêtes, partly, no doubt, because he believed that its silent action would speak to the common people as

---

[3] On the Colisée, see the *Avant-Coureur*, 8 July 1771; Émile Campardon, *Les Spectacles de la foire*, 2 vols. (Paris, 1877), I. 209; Paul Jarry, 'Notes sur le Colisée', *Bulletin de la Société historique des VIIe et XVIIe arrondissements* (1913), pp. 75–84; Alain-Charles Gruber, 'Les "Vauxhalls" parisiens au XVIIIe siècle', *Bulletin de la Société de l'histoire de l'art français, année 1971* (1972), pp. 125–43, especially pp. 132–5.

[4] L. Gachet, *Observations sur les spectacles en général, et en particulier sur le Colisée* (Paris, 1772), pp. 52–3.

[5] [Ducoudray,] *Correspondance dramatique*, 2 vols. (Paris, 1778), I. 69.

[6] Ibid. I. 291; Brazier, *Histoire des petits théâtres de Paris*, new edition, 2 vols. (Paris, 1838), I. 116–17.

[7] See Christel Heybrock, *Jean-Nicolas Servandoni: eine Untersuchung seiner Pariser Bühnenwerke* (Cologne, 1970); Bergman, pp. 71–81.

well as to the cultivated. Diderot called his 'spectacles optiques' 'purement pantomimes' (*A.-T.*, VIII. 463), although they lacked several qualities usually associated with the word. The mute action of characters, in particular, was much less important to Servandoni than the presentation of tableauesque scenes which combined to tell a story, generally borrowed from classical mythology or more recent 'merveilleux' literature. Servandoni had obtained use of the Salle des Machines[8] at the Tuileries in 1738, and from then until 1742, and from 1754 to 1758, showed his spectacles there. The story which each related (usually in seven tableaux) was not necessarily self-explanatory, so that Servandoni provided pamphlets summarizing the plot. The only aural accompaniment to the spectacles was musical, with no verbal explanation forthcoming from the actors themselves. (This fact casts some doubt on Lablée's notion that such works could fruitfully have been performed before the masses in the Revolutionary fêtes.) That a degree of movement was present in Servandoni's tableaux is clear from his own descriptions and from the comments of contemporary witnesses. De La Porte's *Les Spectacles de Paris* of 1756 writes of the latest production, *Les Conquêtes de Thamas-Kouli-Kam*, that it is 'susceptible de beaucoup de mouvements et de coups de théâtre surprenants' (p. 40). But Lablée states that when Servandoni tried to create drama of a pantomimic kind, he failed: 'Ses grands tableaux étaient admirés; mais voulait-il présenter une action dans ses différentes circonstances, il n'intéressait pas. Les bras de ses acteurs, dit Noverre, n'étaient jamais dans l'inaction, cependant ses représentations pantomimes étaient de glace.'[9]

Despite such entertainments as these, the most important creators of pantomime remained the 'little theatres' of fair and boulevard, particularly after the expulsion of the Italian actors from France in 1697. The Parfaict brothers note in their *Mémoires pour servir à l'histoire des spectacles de la foire* that this expulsion opened a vast field to the entrepreneurs of fairground amusements, who regarded themselves as the rightful inheritors of the troupe's repertoire, and enlarged their own companies—usually by hiring actors from the provinces—to satisfy public demand.[10] The success of the forains' efforts, which as

---

[8] This was a small theatre built in 1660 to house the spectacular shows given in honour of Louis XIV's marriage. It continued for many years to be used for court entertainments.

[9] Jacques Lablée, *Du théâtre de la Porte Saint-Martin, de pièces d'un nouveau genre, et de la pantomime* (Paris, 1812), p. 8.

[10] [François and Claude Parfaict,] *Mémoires pour servir à l'histoire des spectacles de la foire, par un auteur forain*, 2 vols. (Paris, 1743), I. 11.

time went on attracted both the 'petit peuple' and the well-to-do, indicates that their entertainments met a need. Audinot, the proprietor of the Ambigu-Comique, had himself acted with the Italians after their return to France, appearing with the troupe in 1762 and 1764. The Parfaicts note that the Italians were so perturbed to find on their return that the forains had taken over much of their old repertoire and audience that they themselves left their theatre in the Hôtel de Bourgogne to set up a rival one at the foire Saint-Laurent, believing, according to the Parfaicts, that rulings which had previously been passed against the forains would hold, and that they would have no competition (I. 237). But in 1724 they realized their mistake, and abandoned their new territory (II. 19).

The connection between the fairs and the Opéra-Comique, to which the Parfaicts refer, was a particularly close one. The *Mémoires pour servir à l'histoire des spectacles de la foire* describes how, before the foire Saint-Germain of 1715, the Sieur and Dame de Saint-Edmé successfully appealed to the administrators of the Académie royale de musique for permission to stage performances of singing and dancing at the fair. (I shall return shortly to the question of the exclusive privileges over the performing arts they professed which were enjoyed for most of the century by the Opéra, the Comédie-Française, and the Comédie-Italienne.) This permission was granted on 26 December 1714. The Edmés' associate, the Dame de Baume, was subsequently granted the same right, and the two spectacles adopted the name of Opéra-Comique (Parfaict, I. 166). Lesage, who was credited with being the inventor of this genre (de La Porte, *Les Spectacles de Paris*, 1754, p. 169), wrote the Edmés a parody of the opera *Télémaque*, performed shortly before by the Académie royale de musique, and this work enjoyed great success (ibid., p. 167). De La Porte refers a year later to the continued good fortunes of the Opéra-Comique, now directed by Monet, and describes the magnificent ballets performed there, as well as commenting on the great size of the audience (*Les Spectacles de Paris*, 1755, p. 98). He emphasizes, as do other observers, the fact that this genre affords a unification of the different arts of poetry, music, and spectacle, with ballet-masters, stage-designers, and 'machinistes' all playing their part. Further on he notes that the dancer and teacher Noverre is having a ballet (*Les Fêtes chinoises*) performed at the same fair of 1755, and in the same theatre (p. 136).

Many of the performances seen at these annual institutions, and later in the boulevard theatres, were offered as isolated spectacles, and not incorporated within the fabric of comic opera or play. The skills of acrobats, tumblers, rope-dancers, and funambulists were all on display at the foire Saint-Germain and the foire Saint-Laurent. Some commentators alluded to the antiquity of such mute performing arts, as they did the art of pantomime itself. Dubos wrote in the *Réflexions critiques* of the *saltatio* of the ancient actors, an aspect of performance which involved all the arts of gesture as well as the movements of dance, and which consequently was appropriate to the most serious moments in tragedy (I. 506 ff.). The anonymous author of the *Lettre écrite à un ami sur les danseurs de corde* observes that a tradition of rope-dancing existed in antiquity, and is referred to by Terence, Horace, Petronius, Juvenal, and Quintilian (p. 4).[11] Some years previously Bourdelot and Bonnet had noted this circumstance,[12] and corroboration is provided by Boulenger de Rivery's discussion in the *Recherches historiques sur quelques anciens spectacles, et particulièrement sur les mimes et sur les pantomimes* of 1751. Some critics were at pains to distinguish such popular performing spectacles as these from the serious craft of acting. A work published in 1790 by De L'Aulnaye, *De la saltation théâtrale*, follows Bourdelot and Bonnet in stressing the fact that the ancients' *saltatio* meant not the vulgar art of tumbling practised by modern entertainers, but the art of gesture in the broadest sense.[13] Even the modern developments of ballet, or what De L'Aulnaye calls the 'art de sauter en mesure, de former avec grâce des pas cadencés' (p. 8), is inadequate to convey the richness of the ancients' skill. De L'Aulnaye's work deals with the part of 'geste' which results from the principle of imitation that it shares with other arts, and 'par laquelle les anciens histrions savaient exprimer toutes les passions, toutes les actions des personnages qu'ils mettaient sur la scène' (p. 2). The Romans so mastered *saltatio*, according to De L'Aulnaye, that they were able to convey through pantomime 'tout ce que renferment et l'histoire, et la fable, et la poésie' (ibid.).

However poor by comparison were the fairground arts of the moderns, they found favour even with those whose wealth and position in society allowed them to be discriminating. So, at least, it

---

[11] The author mentions another authority's enumeration of four types of rope-dancing: one in which the performer spun around the rope, hanging on with feet or neck; another in which he lay on his stomach with arms and legs outstretched, and swooped 'de haut en bas'; a third in which he ran along a rope stretched obliquely; and a fourth in which he walked on a taut rope, and leapt and spun to the sound of flutes (pp. 5–6).

[12] [P. Bourdelot and] Bonnet, *Histoire générale de la danse sacrée et profane* (Paris, 1732).

[13] De L'Aulnaye, *De la saltation théâtrale* (Paris, 1790), p. 101, note 1.

would appear from de La Porte's notice in his 1754 almanac that tumblers, dancers, and acrobats were all hired for the fête at Chantilly in 1722, given in honour of the King by the duc d'Orléans (p. 189). But their place within the tradition of genuinely popular theatre remained. The entr'actes in plays staged at Nicolet's Grands Danseurs du Roi were filled with 'équilibristes', 'tourneuses', and their like. Ducoudray refers dismissively in his *Correspondance dramatique* of 1778 to the social status of such performers, and to the type of audience which found their antics appealing: 'saltimbanques', for him, are 'ces gens de la lie qui font des postures, des sots [*sic*], des tours de passe-passe et d'adresse pour tromper les nigauds' (II. 85).

The development of mute performing arts involving gesture and movement was indirectly encouraged by the prohibitions imposed on unofficial troupes by the protectors of the royal companies. The anxieties felt by those who had, in the previous century, been granted special privileges as titular actors, singers, and dancers to the King have already been mentioned. These fears reached extreme proportions where the activities of the forains were concerned, as Comédiens and members of the Académie royale de musique fought to retain their exclusive right to the performance of French plays and of operas in the capital. The forains displayed ever-increasing ingenuity in their attempts to obey the letter, but not the spirit, of the various royal ordinances to which the acting and singing establishments appealed. The Parfaicts list a number of the decrees which aimed to protect the monopoly of the King's performers. In 1699 the lieutenant-general of police, d'Argenson, pronounced that no private individual should perform any 'comédie' or farce, but the foire proprietors appealed against the damages he awarded the Comédiens Français, and carried on as before (I. 18). In June 1703, during the foire Saint-Laurent, the parlement of Paris confirmed the two edicts; the forains 'officially' complied by performing only detached scenes from plays, complete in themselves, and by increasing considerably the proportion of 'jeux' in their spectacles (ibid., pp. 31–2). In 1706 d'Argenson pronounced two new sentences against the fairground entrepreneurs, forbidding them to stage any play which contained dialogue and which was contrary to the rules of propriety and decency (pp. 47–8). At about this time the Comédiens Français protested their weariness at having to lodge complaints about the forains twice yearly, during the seasons of the foire Saint-Laurent and the foire Saint-Germain respectively (p. 69). In 1708 the fairground proprietors Alard and widow Maurice made an attempt to reach a legal agreement with the Opéra: anticipating a fresh judgement against them, they appealed to the director of the Académie royale de musique, Guyenet, for permission to use changes of décor in their theatres, and to employ singers for divertissements and dancers for ballets. (Such permission was in the gift of Guyenet by virtue of the letters-patent granted to the Opéra the previous century.) Alard and Maurice were accorded the favour they asked, and were thus able to protect themselves against the edicts of the parlement (pp. 73–4).

Not all the forains were prepared to deal with the authorities, however. When the 'arrêt' of 22 February 1707 forbade the use of dialogue by non-official troupes, they substituted monologues, while the non-speaking actors communicated through signs (Parfaict, I. 59). The same year the Comédiens Français complained that the troupe of de la Place and Dolet employed a kind of dialogue in which actors swiftly alternated with one another on stage, spoke freely to each other in the wings, or repeated loudly what another character on stage had whispered (pp. 63–4). D'Argenson's response to this complaint was to order that the offending parties' theatre be demolished. Later, in 1709, the forains performed parodies, imitating well-known actors of the Comédie-Française in gesture and voice, and pronouncing meaningless words in the form of alexandrines, and in tragic tones. When Guyenet told Alard that permission to breach the 'privilèges' of the Académie royale de musique was being withdrawn, Alard obediently staged silent plays; but, as the *Mémoires pour servir à l'histoire des spectacles de la foire* reports,

comme le public s'était plaint à la précédente Foire de l'obscurité de beaucoup d'endroits de ces pièces, causée par l'impossibilité où les acteurs étaient d'exprimer par des gestes des choses qui n'en étaient pas susceptibles: on imagina l'usage des cartons, sur lesquels on imprima en gros caractères, et en prose très laconique, tout ce que le jeu des acteurs ne pouvait rendre. (I. 109–10)

These written scrolls were unfurled before the audience by each actor at the moment of his supposed speech. Following this was the device of substituting for prose dialogue couplets set to known airs (called vaudevilles), to be sung by the audience. To ease performance by the spectators, the orchestra first played the air, members of the audience in the stalls and pit who had been hired by the proprietors for this purpose sang the words, and the rest followed suit. An alternative was introduced in 1712, when 'écriteaux' were dangled from the flies of the theatre: the audience found these easier to read, and the actors were freed to perform their 'jeux de théâtre' (I. 137).

Bachaumont's *Mémoires secrets* reports on 2 December 1771 that Audinot is plagued by the jealousy of his rivals in the official theatre, and has been reduced by decree to being no more than a purveyor of popular spectacles, prohibited from having dances performed on his stage or using the greater part of his orchestra. Bachaumont's continuator returns to this theme in the entry for 21 May 1785, which recalls that when Audinot first set up his theatre, the Opéra, Comédie-Française, and Comédie-Italienne had each imposed their separate vetoes on the activity of his troupe (XXIX. 35 ff.). Audinot's response had been to use marionnettes instead of real actors, and later he obtained permission to replace them with live children. In 1786 the almanac *Les Petits Spectacles de Paris* reports on the troupe of the comte de Beaujolais, the Petits Comédiens. The entrepreneurs of the spectacle mounted by this troupe have found a way of delighting their public with a kind of pantomime never before seen in France:

> ne pouvant faire chanter sur leur théâtre, ils ont des chanteurs et des acteurs cachés dans la coulisse, en sorte que les enfants qui sont sur la scène n'ont que des gestes à faire. Mais cette gesticulation ou pantomime s'exécute avec tant d'art, elle est tellement d'accord avec les paroles et le chant, que l'illusion est complète, et qu'il semble que les acteurs qu'on a sous les yeux n'aient point d'interprètes.[14]

The Romans, this commentator observes, used to have one actor executing movements and a partner declaiming, but the Beaujolais troupe's practice seems a new variation.

The persecutions suffered at a later date by the former actor Valcour, stage manager at the Théâtre des Délassements Comiques in the boulevard du Temple, are described by the authors of the *Almanach général de tous les spectacles de Paris et des provinces* for 1791. Before the Revolution the police, encouraged by directors of other theatres in the boulevard, decided to harass Valcour and his director Colon: 'on n'y pouvait faire paraître que trois acteurs à la fois; on n'y parlait pas, on n'y jouait que la pantomime, et une gaze séparait le

---

[14] *Les Petits Spectacles de Paris* (Paris, 1786), p. 183. Baron's *Lettres et entretiens sur la danse ancienne, moderne, religieuse, civile et théâtrale* (Paris, 1824) mentions that a similar procedure was followed, at Noverre's suggestion, in the performance of an opera in Vienna (p. 213).

public de la scène.'[15] But on 14 July 1789 Valcour tore down the gauze with a cry of 'Vive la liberté!'

Not all the actors performing at the fairground were beings of flesh and blood. Puppet shows provide one example of the successful evasion of such penalties as had been decreed for those who infringed the monopoly of the official performing troupes. But marionnettes had been the earliest fairground entertainers, according to the Parfaicts, and they were thus well established before the forains had recourse to them as a way of dodging prohibitions on acting. One commentator claims that puppets were the only performers at the fair not to suffer the persecution of the Premiers Gentilshommes de la Chambre, and able freely to play the entire repertory of old and new dramas.[16] Two reports of 30 August 1707 and 3 August 1708 reveal that entrepreneurs like Alard and Maurice succeeded in having comedies and comic operas performed at the fair by preceding them with puppet shows or displays by rope-dancers, and thus deflecting official attention from the work that followed.[17] Their practice was officially condemned by the parlement on 2 January 1709, however, and the offending parties were enjoined to use their theatres only for the type of spectacle for which they had originally been licensed. In 1719, when an official edict closed all the theatres of the fair, performances by puppets and rope-dancers were allowed to continue. The following year a transaction took place between the Comédie-Française, the Comédie-Italienne, and Opéra, on the one hand, and the 'petits théâtres' on the other: the latter were allowed to perform plays in which there was some dialogue and song, and puppets were still allowed to say, sing, and enact whatever they wanted (Magnin, p. 153).

The three playwrights Lesage, Fuzelier, and Dorneval, who wrote for the fairs, themselves became puppeteers at de la Place's and Dolet's Théâtre des Marionnettes, after the jealousy of the Comédiens Français had led to the closure of the theatre (owned by Francisque) for which they had worked. According to the Parfaicts, their shows were very successful (II. 4). The playwright Piron, author of *Arlequin-Deucalion* (first performed at the foire Saint-Germain on 25 February 1722, in Francisque's theatre), had previously worked only for

[15] Charles-Guillaume Étienne et al., *Almanach général de tous les spectacles de Paris et des provinces*, 2 vols. (Paris, 1791–2), I. 211.

[16] [De La Porte,] *Les Spectacles de Paris* (1754), p. 185; also *Les Spectacles de Paris, ou Calendrier historique et chronologique des théâtres* (Paris, 1792), part 2, p. 34.

[17] See Charles Magnin, *Histoire des marionnettes en Europe*, 2nd edition (Paris, 1862), p. 152.

puppeteers. The play itself reflects this experience and the climate of the times during which it was written and performed. Arlequin-Deucalion searches all over Parnassus for material with which to create man and eventually lays his hands on a wooden puppet, which immediately begins to utter its nonsensical speech.

Some lines from Lemierre's poem *Les Fastes* suggest the popularity of marionnettes in mid-century, and the appeal of their 'pantomime':

> Jolis petits acteurs, taillés en plein coffret,
> Trottant, gesticulant, le tout par artifices,
> Tirant leur jeu d'un fil, et leur voix des coulisses:
> Point soufflés, point sifflés, comme il arrive ailleurs:
> Nulle tracasserie, encore moins de noirceurs,
> Ouvrant leur jeu, jamais de rhume sur l'affiche.
> (Quoted by Ducoudray, *Correspondance dramatique*, II. 84)

Both Nicolet and Audinot, as might have been expected, drew in crowds with their puppet shows. The former had had a theatre for marionnettes at the foire Saint-Germain of 1753, before he set up in the boulevard du Temple. Audinot presented similar entertainments at the same fair from 1769, using large puppets which were modelled after his former colleagues at the Opéra-Comique. His pantomime *Les Comédiens de bois* was said by Bachaumont in his entry for 16 February 1769 to be attracting all Paris. The devotion to such spectacles shown by many privileged members of society is widely attested. The duchesse du Maine, who did much to aid the rebirth of pantomime in eighteenth-century France, often received puppeteers at Sceaux.[18] Voltaire is described by Mme de Graffigny as having organized spectacles in which puppets played their part (Magnin, p. 196). Mlle Pellissier, an actress with the Opéra, was allegedly so devoted to Polichinelle, dame Gigogne, and other characters of the puppet stage that she paid a puppeteer to provide her with a show twice daily.[19]

The pursuit of illusion, if not an object with most puppeteers, does seem to have been marked in one or two cases. This is unsurprising, perhaps, in view of the fact that many puppeteers had adopted their 'comédiens de bois' only as a substitute for the real actors they were forbidden to use. Some puppets were nearly life-size, like the Polichinelle shown by Pierre Second at the foire Saint-Germain in 1762.

---

[18] See Marie-Françoise Christout, *Le Merveilleux et le 'théâtre du silence'* (The Hague and Paris, 1965), p. 132.

[19] *Les Spectacles des foires et des boulevards de Paris* (Paris, 1778), p. 25.

The Italian Carlo Perico set up a puppet theatre on the boulevard du Temple in 1778, and Mme Vigée-Lebrun notes in her memoirs that his puppets were so well-made as sometimes to induce the illusion that they were real human beings.[20]

The troupe of the comte de Beaujolais was of 'petits comédiens *de bois*' when it began performances at the Palais-Royal in 1784, but from the following year onward the puppets were replaced by children. As already noted, this was also done at the Ambigu-Comique in 1773, a fact which led to confusion in some members of the audience. The *Almanach forain* of 1773, by the fairground author Arnould-Mussot, reports the following conversation between two women emerging from a performance at the latter theatre:

> Ma bonne amie, je ne regrette pas mes douze sols; ceci mérite d'être vu: ma foi, ces petits comédiens de bois sont tout à fait charmants.—Oui, lui répondit l'autre, on jurerait qu'ils sont vivants: les fils qui servent à les conduire sont ménagés avec tant d'adresse qu'ils sont réellement imperceptibles.—Oh! j'ai donc l'œil plus fin que vous, ajouta la première; car après les avoir fixés quelque temps avec beaucoup d'attention, j'ai aperçu très difficilement les fils qui les font mouvoir, et particulièrement ceux du petit Arlequin.

But children had been substituted for puppets two months previously (Arnould-Mussot, op. cit.). The *Almanach général de tous les spectacles de Paris et des provinces* for 1791 reports that the 'fantoccini chinois' set up next to the Ambigu-Comique in the boulevard du Temple were a troupe of puppets, sometimes replaced by live actors (I. 266). The fact that marionnettes were still popular after restrictions on theatre performance were lifted with the Revolution shows that they had an appeal beyond that engendered by the repressive circumstances of earlier years. But there is no doubt that a part of their popularity derived from the spur to developing the puppeteer's art which such repression had provided.

Another spectacle which presented a type of pantomime was that of shadow-puppets, or 'ombres chinoises'. Their appeal seems to have been predominantly to the upper and middle classes, although the entrepreneur Ambroise did show his *Théâtre des récréations de la Chine*, which ended with shadow-puppets, at the fairs of 1775 and 1777. (The *Almanach forain* incidentally notes that this spectacle was one to which 'les ecclésiastiques pouvaient venir sans scrupules'.) A witness writing of the Théâtre Séraphin, which set up in 1775 and won great

---

[20] Jacques Chesnais, *Histoire générale des marionnettes* (Paris, 1947), pp. 123–4.

popularity for this type of entertainment, stated that the 'petit peuple' was not much in evidence there (Chesnais, p. 129). The bourgeoisie and the people of quality, on the other hand, were eager spectators. The *Correspondance littéraire* of 15 August 1770 refers to the mention in Cléry's memoirs of the Queen summoning Séraphin to give three performances every week during the carnival season. The Court, Cléry adds, was well pleased, and Séraphin then asked the King for permission (which was granted) to set up a theatre in Paris without having to pay the usual dues to the official establishments.[21]

This form of entertainment had a long tradition. Before Séraphin opened his spectacle in 1774, Audinot had presented shadow puppets at the foire Saint-Germain in 1760; but Gueullette mentions that earlier examples had existed in France.[22] The *Guide des amateurs à Paris* notes that Séraphin's puppets performed some of the activities long associated with the fairs:

> Les ombres chinoises produites par différentes combinaisons de lumières et d'ombres y représentent au naturel toutes les attitudes de l'homme et y exécutent des danses de corde et de caractère d'une précision étonnante. Des animaux de toute espèce y passent en revue et font aussi tous les mouvements qui leur sont propres, sans qu'on aperçoive ni fil, ni cordon pour les soutenir ou les diriger. (Campardon, II. 393)

But the plays were not, as this might seem to suggest, silent pantomimes: dialogues in printed form exist along with drafts for action. Séraphin showed several 'pièces de circonstance'. In 1789 he staged *L'Apothicaire patriote*, in which the courage and patriotism of the women who had gone to Versailles to bring the royal family back to Paris is celebrated. This was followed by a piece entitled *La Démonseigneurisation*, followed in turn by *La Fédération nationale*, titles which suggest that Séraphin's shadow-plays were not aimed primarily at an audience of children (*Feu Séraphin*, p. 12). The question of their repertoire coloured the judgement which many commentators passed on the proprietors of boulevard and fairground theatres. As the details of Séraphin's 'pièces de circonstance' reveal, the entrepreneurs of aesthetically limited spectacles such as shadow-plays were anxious to follow the lead of the regular theatres in treating matters of the moment. Naturally the political pretensions of such small entertain-

---

[21] See *Feu Séraphin: Histoire de ce spectacle 1776–1870* (Lyon, 1875), pp. 2–3, footnote 4.

[22] Georg Jacob, *Geschichte des Schattentheaters*, 2nd edition (Hanover, 1925), p. 186.

ments were comparatively slight; but the proprietors of larger concerns, like Nicolet and Audinot, were at least theoretically in a position to provide the 'royal' theatres with competition once restrictions on playhouses had been lifted.

For some commentators, the nature of the acrobatic and pantomimic skills fostered by the fair meant that they made limited and even reprehensible forms of entertainment. Ducoudray writes in 1779 that Nicolet 'ne ser[t] qu'à perpétuer le mauvais goût, le genre obscène, et qui n'exist[e] qu'à la honte des mœurs et des lumières du dix-huitième siècle.'[23] Gachet subscribes to the same view, although he states it in less extreme form. For him, Nicolet's rope-dancers and acrobats would deserve more attention than they were granted by serious men 'si on pouvait les [i.e. performances] diriger vers quelque utilité publique; mais je ne dissimulerai pas qu'il s'y passe encore quelquefois des indécences' (pp. 19–20). Moral disapproval of non-dramatic spectacles at the fair and in the boulevard, especially the kinds of gymnastic display that were commonly offered, was sometimes expressed in terms of the human risks in which they involved their performers. The latter allegedly debased their audience's sense of the value of life by seeming so careless of their own. One commentator recalls in 1777 that this objection had been voiced by Mercier in his work *L'An 2440* (*Les Spectacles des foires et des boulevards de Paris*, p. 143). The almanac *Les Spectacles de Paris* for 1792 writes of the 'sauteurs, lutteurs et voltigeurs' of Nicolet's Gaîté: 'Nous ne parlerons de ces différents exercices que pour dire qu'ils nous font éprouver de vives inquiétudes pour ceux qui y montrent le plus de talent, et dont la vie est sans cesse exposée' (p. 24).[24]

Dumont finds Audinot's Ambigu-Comique to be the only boulevard theatre where a little taste is in evidence.[25] Audinot is admittedly 'livré tout entier à ce qui peut flatter le goût du public', but this does not signify that he panders to its baser instincts. Indeed, according to Bachaumont's entry for 1 January 1785, he is the father of

---

[23] [Ducoudray,] *Il est temps de parler, et Il est temps de se taire*, 2 vols. in 1 (Paris, 1779): *Il est temps de se taire*, pp. 11–12.

[24] See also J. B. de La Salle, *Les Règles de la bienséance et de la civilité chrétienne, à l'usage des écoles chrétiennes des garçons* (Reims, 1736), p. 133: 'L'honnêteté ne permet pas . . . de se trouver aux spectacles des danseurs de corde, qui exposant tous les jours leurs vies, aussi bien que leurs âmes, pour divertir les autres, ne peuvent être ni admirés ni même regardés par une personne raisonnable, puisqu'ils font ce qui doit être condamné de tout le monde, en suivant les seules lumières de la raison.'

[25] *Le Désœuvré mis en œuvre*, in [Nougaret,] *La Littérature renversée, ou L'Art de faire des pièces de théâtre sans paroles* (Paris, 1775), p. 15.

'decent' pantomime, and it is thanks to him that the boulevard du Temple has virtually become a school for morality. A *Lettre sur les pantomimes* written (probably by Nougaret) ten years previously had emphasized its author's belief that pantomime should depict 'les mœurs', by which he clearly means a preponderance of 'bonnes mœurs';[26] and this idea is found again in de L'Aulnaye's later work *De la saltation théâtrale*, which recalls that the most respectable pantomime artistes in ancient Greece were called ethologists, or 'peintres de mœurs' (p. 28).

Pantomime as staged in the boulevards was often seen as exemplifying the union of different art-forms, and in this respect could be regarded as a more serious entertainment than its critics acknowledged. The boulevard theatres developed it far beyond the often primitive bodily arts of the fair. The *Almanach général* of 1791 greeted Audinot's success in staging pantomime with acclaim. According to its authors, 'nulle part on n'approche de l'ensemble et de la perfection avec laquelle on les [pantomimes] exécute chez Audinot' (I. 172). No effort or expense was spared in the mounting of these spectacles, which brought together 'musique charmante, jolis ballets, décorations fraîches, tableaux séduisants, costumes agréables, etc.' (p. 179). In 1761 Dumont remarked less admiringly of Nicolet's pantomimes that they too were mixed spectacles 'où se trouve réunis l'extrêmement merveilleux et l'extrêmement ridicule' (*Le Désœuvré mis en œuvre*, p. 23).

The difficulty of pleasing the public with pantomime in its purest wordless state was acknowledged by a number of writers in the course of the eighteenth century. The Parfaicts' observation about the 'impossibilité où les acteurs [in Alard's troupe] étaient d'exprimer par des gestes des choses qui n'en étaient pas susceptibles' has already been noted. Diderot's view, later on, was that pantomime worked most effectively when supported by speech. In the *Entretiens sur 'le Fils naturel'* Dorval calls for a playwright of genius to combine pantomime with discourse, to follow a scene containing dialogue with a silent one, and to take advantage of this succession, 'surtout de l'approche ou terrible ou comique de cette réunion qui se ferait toujours' (p. 115). The Moi-character recalls that in ancient spectacles music, declamation, and pantomime were sometimes brought together, and sometimes kept separate, but is told by Dorval that the effect he himself seeks is only occasionally equivalent to the ancient model (ibid.). The illustration Dorval proceeds to give of the uniting of pantomime

---

[26] In *La Littérature renversée*, p. 30.

and speech in a play (of his own invention) reveals nothing so much as Diderot's taste for melodrama, presenting tableaux of a father alone with the corpse of his son, and the mother performing acts of piety before losing consciousness as she perceives the body (pp. 116-17). Scenes like this, according to Dorval, make for tragedy; but he says that such dramatic presentation would require particular authors, actors, theatres, and even audiences for its full effect to be realized. To Moi's nervous observation that Dorval's imagined scene calls for 'un lit de repos, une mère, un père endormis, un crucifix, un cadavre, deux scènes alternativement muettes et parlées'—a conflation of properties and personalities which the classical mould of French tragic drama would certainly have prohibited—Dorval can only respond: 'Ah! bienséances cruelles, que vous rendez les ouvrages décents et petits!' (p. 117).

Other writers on the subject shared Diderot's opinion that pantomime combined happily with speech, however attracted they may have been to the idea that the former was a kind of language which transcended the barriers erected by divergences of tongue. Although Dorat's section on pantomime in *La Déclamation théâtrale* affirms the belief that every sentiment and passion may be conveyed through gesture, the assumption behind the whole poem is that this mute language should simply support and intensify the thoughts and impressions conveyed in drama by words. But none of this is to deny what most critics from Diderot onwards readily accepted, that there are moments in the performance of any play when speech should give way to silence, and to the conveying of emotion through facial expression and bodily movement. In the *Lettre sur les sourds et muets*, discussing the communication of thought by language, Diderot states his belief that 'il y a des gestes sublimes que toute l'éloquence oratoire ne rendra jamais' (p. 47). He gives two illustrations of a silent *actio* that no verbal expression could match, still less surpass. The first is of Lady Macbeth walking silently and with closed eyes, 'imitant l'action d'une personne qui se lave les mains, comme si les siennes eussent encore été teintes du sang de son Roi qu'elle avait égorgé il y avait plus de vingt ans'. There is nothing, Diderot writes, so full of pathos as this movement of the hands and this silence. Diderot's second example is of a woman announcing to her distant husband the death of their son. Having drawn close to the tower where the former is imprisoned, she takes a handful of earth and scatters it in the form of a cross on the child's body. Her husband understands the sign, and starves himself to death. Both these examples, according to Diderot, reveal a force of expression which the energy of spoken language cannot touch. 'On oublie la pensée la plus sublime; mais ces traits ne s'effacent point' (p. 48).

Gachet's *Observations sur les spectacles en général, et en particulier sur le Còlisée* conveys his admiration for Audinot's art of pantomime, described as displaying more finesse than Nicolet's. But the author still reveals reservations about the possibilities which the genre offers: 'malgré tout, je regarde toute pantomime comme un jeu imparfait, tant que nous n'aurons point attaché à certains gestes précis des idées précises' (p. 20). The difficulty is that many passions find expression in gestures which are undifferentiated, or barely differentiated from one another.

Il est des passions qui emploient des gestes communs ou ressemblants, comme par exemple le mépris, la haine, l'aversion, la colère, le dépit, le désespoir: ceux dont peuvent se servir l'espérance, la joie, l'amitié, la reconnaissance, l'amour, ont aussi entr'eux beaucoup d'affinité et de ressemblance. Il en est même de communs à des passions opposées: la vengeance, la bienfaisance satisfaites, exprimeront par les mêmes signes leur contentement. (pp. 20–1)

The consequence is that an audience, lacking specific guidance, may interpret the sense of a movement wrongly: 'L'esprit du spectateur erre donc au gré de son imagination, qui de moment à autre a la mortification de reconnaître son erreur, en voyant qu'elle est entrée dans des pensées souvent différentes de celles qu'elle croyait deviner' (p. 21). Gachet goes on to examine the attempt to remedy these deficiencies recently made at the Colisée, where the audience had been supplied with a programme alerting it to the actions that were to follow. But this expedient could not succeed, he reasons, because the pantomime actor cannot be bound by such prescriptions, unless Riccoboni's theories about 'method' acting be taken to their extreme limits. It is simpler, Gachet objects, to recognize that words provide the easiest, most exact, and most natural means of expression in the theatre, when their use is not prohibited by the checks and obstacles which the actor outside the official Paris theatres had to face. Even if the silent performer succeeded in making his audience as involved in his thoughts and impressions as would his speaking counterpart, there could be only gain in the alliance of pantomime and words: 'la voix et ses différentes inflexions doublent notre sensation et par conséquent notre plaisir' (p. 22).

Gachet's conclusion, pointing as it does to the limitations of *actio* in theatrical performance, serves as a fitting epigraph to the varied programme of pantomimic arts which the eighteenth century saw, and is fully in accord with the beliefs about silent expressiveness elaborated by earlier writers. Limited though it may have been in its pure form, pantomime could greatly intensify the effect of an actor's words, as Diderot and others saw. Diderot's remarks in the *Entretiens sur 'Le Fils naturel'* about the linking of pantomimic action with verbal discourse reach similar conclusions to those drawn by other writers on the eloquence of the pulpit, courtroom, and political assembly, which I discussed in chapter 2. The persecution of unofficial theatres by the royal companies, which ended only with the Revolution, undoubtedly helped French acting to develop its tradition of bodily eloquence, and it has retained that tradition to this day. Unquestionably, too, the interest in pantomime shown not just by theorists like Dubos and Diderot, but also by a public of well-to-do citizens, aristocrats, and even royal patrons, helped further this development. An element of pantomimic action was introduced by some of the Comédiens Français, such as Lekain, into their own stage performance, although they may have been readier to display this facet of their art in private theatricals like Voltaire's than in the official theatre. As with discussions of *actio* outside the playhouse—in the church or the court of law—it was possible for serious supporters of pantomime to argue its ancient origins and hence its respectability. The same approach was sometimes adopted in writings on dance in the eighteenth century, and I shall describe some of its implications in the next chapter. In all these cases, the dignity of classical precedent was seen as adding prestige to more recent versions of bodily communication, but in every instance it was clear to contemporaries that the appeal of such art to modern sensibilities, including that of the uneducated masses, was crucial to its success.

CHAPTER FIVE

'Bodily Eloquence' and Dance

Many works were published in eighteenth-century France about the history of dance, the description and teaching of it, its usefulness in different walks of life, and its place in theatrical spectacle. The question whether it might fruitfully be combined with other arts, such as singing and acting, was frequently debated in connection with opera and the merits of the French type in comparison with the Italian. Dance was regularly described as another of the mute and universally comprehensible languages of gesture and movement, like pantomime, and sometimes (as in the discussions of the ancients' *saltatio*) it was claimed to be its equivalent: eighteenth-century writings on these topics refer to Pylades and Bathyllus as dancers as often as they call them pantomime artistes. It was variously argued that there were links between dance and the acting style of conventional drama. Noverre, whose *Lettres sur la danse* was the most influential work on the subject published in the eighteenth century, explicitly aligned his ideas about 'action ballet' with Diderot's theory of *drame*, and many of Noverre's observations echo discussions of bodily acting in the *Entretiens sur 'Le Fils naturel'* and *De la poésie dramatique*.

The origins of dance were generally seen to lie in the pre-civilized 'natural' state, and its movements said to correspond to the spontaneous expression of emotions. The precise nature of the relationship between emotions and the actions of dance was not investigated, although the notion that the basic movements of dance reflect certain invariable bodily responses to psychic stimuli was posited. The *Encyclopédie* article 'danse' relates another physical means of expression which it likewise sees as spontaneous—that of song—to the 'geste' of dance, and argues that the former leads naturally to the latter:

Le chant si naturel à l'homme, en se développant, a inspiré aux autres hommes qui en ont été frappés des gestes relatifs aux différents sons dont ce chant était composé; le corps alors s'est agité, les bras se sont ouverts ou fermés, les pieds ont formé des pas lents ou rapides, les traits du visage ont participé à ces mouvements divers, tout le corps a répondu par des positions, des ébranlements, des attitudes aux sons dont l'oreille était affectée: ainsi le chant qui était l'expression d'un sentiment a fait développer une seconde expression qui était dans l'homme ce qu'on a nommé danse.

The author of the article 'courante' in a later encyclopaedia, Panckoucke's *Encyclopédie méthodique*, similarly affirms that dance has been from its origin a naïve expression of men's sensations. Every dance, it continues, should therefore convey some affection of the soul. If it fails to do so, it loses its original character and is merely an 'abus de l'art'. This condition is even more necessary when dance is transferred to the theatre, 'parce que la représentation fait le caractère essentiel et distinctif de l'art dramatique dont elle fait alors partie'.

Eighteenth-century historians of dance frequently observe that the art is discussed by both Aristotle and Plato, and mention the part it played in the performance of Greek tragedy and the enactment of religious ceremonies. Ménestrier, the author of a work on ancient and modern ballet (1682), had written that in modern Spain and Portugal it still performed this role in churches and the most serious kind of procession, and noted that the performance of ballets had been an important aspect of religious activity for Jews and Christians as well as for pagans.[1] The author of a *Lettre critique sur notre danse théâtrale* (1771) ventures the opinion that the dance of the ancients bore a faint resemblance to another modern but non-religious type of entertainment: 'On peut (s'il est permis de comparer les petites choses aux grandes) prendre une légère idée de ce genre de spectacle en voyant les pantomimes dansantes que le sieur Audinot fait exécuter sur son théâtre des boulevards.'[2]

As the Jesuit Ménestrier's interest in the art of dance might suggest, the performance of ballets was included in the spectacles mounted by Jesuit colleges. Baron's *Lettres et entretiens sur la danse ancienne, moderne, religieuse, civile et théâtrale* (1824), which drew heavily on Cahusac's *La Danse ancienne et moderne* (1754), describes how dance has advanced from the crudeness of its early state as an unrefined manifestation of inner feelings like pleasure or pain, anger or tenderness, affliction or joy, to be a regulated art governed by the principle of imitation, so purified that it can properly be performed on the Jesuit stage (pp. 9–10). The Jesuit Père Le Jay, who in 1725 published a treatise on ballet to follow Ménestrier's, wrote that 'Le ballet est une danse dramatique qui montre, d'une manière agréable et faite pour plaire,

---

[1] [Ménestrier,] *Des ballets anciens et modernes* (Paris, 1682), preface.
[2] *Lettre critique sur notre danse théâtrale* (Paris, 1771), p. 14, note 6.

les actions de toute espèce, les mœurs et les passions, au moyen de figures, de mouvements, de gestes et à l'aide de chants, de machines et de tout l'appareil théâtral.'[3] The entry for 'ballets de collège' in the *Encyclopédie méthodique* notes that at the Jesuit college Louis-le-Grand a tragedy and a 'grand ballet' were performed every year, the ballet usually serving as an intermezzo within the play. To the Jesuits is also due the Christianization of ballet through its inclusion within operas whose libretti, like the texts of the Latin tragedies the pupils performed, were used as a vehicle for religious propaganda. (This particular movement reached its apogee in the operatic work of Marc-Antoine Charpentier.)[4] In 1697 the pupils at Louis-le-Grand took relief from performing Le Jay's severe tragedy *Posthumius dictator* by dancing the Ballet de jeunesse composed by the ballet-master and choreographer Beauchamps (ibid., p. 207). Here, as often in such productions, they were professionally helped by artistes from the Opéra.

Rémond de Saint-Mard's *Réflexions sur l'opéra* (1741), which comments disapprovingly on the ubiquity of dance in Paris—'Ne trouvez-vous pas . . . qu'on donne trop d'étendue à la danse?'[5]— acknowledges that few will echo his complaints. The French have acquired such a taste for this art that 'nous faisons des vœux pour voir finir le plus bel air du monde, dès que nous imaginons qu'il peut être suivi d'un ballet' (pp. 92–3). He himself, he adds, is more touched by a well-executed and agreeable air than by an 'entrée de ballet'. While his fellows enjoy only visual pleasures, he prefers the joy of music, which moves his heart (p. 93). But that dance itself, like other arts perceived visually, might powerfully affect the heart was firmly stated by many commentators; and this notion was central to the argument that it was an imitative spectacle, as we shall see. At all events, its popularity was such that one established theatre sought permission to include dance among its tragic and comic offerings. The 'privilège' granted to the Opéra by Louis XIV meant that no other theatre, in theory, might include dance among its spectacles. (There were some

---

[3] See Pierre Peyronnet, 'Le Théâtre d'éducation des Jésuites', *Dix-huitième Siècle*, 8 (1976), 107–20. Pantomimes had earlier featured in court entertainments, especially those of Molière, where their purpose was less serious: they formed part of the musical–dramatic *Gesamtkunstwerk*. They were not, however, present in tragedies, as they were in Jesuit productions.

[4] See Robert Lowe, 'Les Représentations en musique au collège Louis-le-Grand de Paris (1689–1762)', *Revue d'histoire du théâtre*, 11 (1959), 205.

[5] [Rémond de Saint-Mard,] *Réflexions sur l'opéra* (The Hague, 1741), p. 55.

exceptions to this rule, as illustrated by the performances of Molière's comédies-ballets, and the sanctioning of certain entertainments involving dance at the fairs.) The Comédie-Française, however, wanted a greater freedom in this respect than the Opéra was prepared to grant it, as a poem entitled *Remontrances de MM. les Comédiens Français au Roi pour obtenir de Sa Majesté la suppression d'un arrêt du Conseil qui leur défend les ballets* (1753) reveals:

> Sire, vos fidèles sujets,
> Les gens tenant la comédie,
> Paisibles suppôts de Thalie,
> Et tous ennemis de procès,
> Osent se plaindre du succès
> De cette fière Académie
> Par qui leur troupe est avilie,
> Et voit proscrire ses ballets.[6]

By contrast, the author inaccurately claims, the temple of Thalia, the muse of comedy, is open

> A tous les sauteurs d'Italie.
> Or admirez, Sire, avec nous
> Ce que doit l'Europe et la France
> A cette Italienne engeance. (p. 2)

Meanwhile, the public is bored by the Comédie-Française's traditional performances of Corneille's and Racine's tragedies, and has taken its custom away to 'admirer nos bateleurs/Ainsi recrutés par la foire'. The Comédie-Française, in short, wants to be allowed to win back its audiences by introducing something akin to the gymnastic spectacles of the forains, as well as ballets, into their own productions. But the King's Chancellor refuses them permission to do so:

> Contre nous seuls il est sévère,
> Et veut proscrire pour jamais
> Et nos sauteurs et nos ballets. (p.9)

After this appeal, the Comédiens Français nevertheless resumed the performance of dance they had already inaugurated, but the jealousy of the Opéra led to the imposition on them of highly restrictive conditions: their orchestra was to employ only six violins, and the number

---

[6] *Remontrances de MM. les Comédiens Français au Roi pour obtenir de Sa Majesté la suppression d'un arrêt du Conseil qui leur défend les ballets sous peine de 10,000 livres d'amende* [1753], p. 1.

of dancers and singers they might use was also severely limited. After 1792, when they might legitimately have performed as many ballets or balletic interludes as they desired, the actors of the Théâtre de la Nation (previously the Comédie-Française) gave up dance altogether.[7]

The art of opera had close links with dance, whether the former was the 'grand' kind associated with Lully and Quinault, or the light opera which had grown up in the fairs or been developed by the Italians as opera buffa. The extent of the quarrel between the supporters of grand opera and the devotees of the Italian model, which I shall not discuss here, shows the interest which opera in general aroused in the eighteenth century. Diderot's review of the anonymous brochure called *Pantomime dramatique*, in which he makes the remarks already noted about Servandoni's optical spectacles, comments on the influence of the 'misérables bouffons' who came to Paris from Italy in 1751, and who taught the French that a cry or gesture often illuminated a situation more directly than the long speeches and recitative of grand opera (p. 458).[8] A later writer, Quatremère de Quincy, confirmed in 1789 that in opera buffa gesture frequently appeared more important to performers and spectators than did music. He added that Italian serious opera was different, offering virtually nothing to the eye; 'on n'y parle à l'âme que par les sons et l'organe qui les transmet.'[9] But the French public's taste for visual spectacle has meant, according to Quatremère, that it

> prend assez souvent le change, et qu'il y règne une sorte d'équivoque du plaisir. Ne prend-on pas souvent l'expression du jeu pour l'expression du chant, l'esprit du poète pour le talent du musicien, l'intérêt dramatique pour l'intérêt musical? L'acteur n'y reçoit-il pas plus de *Bravo* que le chanteur? Je n'en sais rien; mais j'entends toujours louer les chanteurs sur leur jeu, applaudir la scène à la place de l'air, et le spectacle pour la musique; on dirait que l'acteur chante pour les gestes, et que le peuple n'écoute qu'avec les yeux. (Ibid.)

The liveliness of the Italians, in music as well as acting, was what had caused the disaffection of many French spectators with their own

---

[7] See Marion Hannah Winter, *The Pre-Romantic Ballet* (London, 1974), p. 31. See also Henri Lagrave, *Le Théâtre et le public à Paris de 1715 à 1750* (Paris, 1972), p. 367, and Sylvie Chevalley, 'Les Bals de la saison d'hiver en 1716–1717', *Comédie-Française*, 66 (1978).

[8] See also Eugen Hirschberg, 'Die Encyklopädisten und die französische Oper im 18. Jahrhundert', D.Phil. (Leipzig, 1903), p. 104.

[9] Quatremère de Quincy, *Dissertation sur les opéras bouffons italiens* (Paris, 1789), p. 10.

operas. Lully's compositions were described in Noverre's *Lettres sur la danse* as cold,[10] and in 1787 Compan's *Dictionnaire de danse* repeats the charge, reasonably equating the style of Lully's music with the style of dance it accompanied. In Lully's day, he writes, music and dance in general were monotonous, frigid, and without individuality.[11] Quinault's libretti were variously judged. Diderot expresses two different views of them: one, voiced by Lui in *Le Neveu de Rameau*, that they resembled Lully's music in their lack of warmth, and that to make them into operas was like setting La Rochefoucauld or Pascal to music (p. 86); and the other, voiced by Dorval in the *Entretiens sur 'Le Fils naturel'*, that Quinault might be read with the keenest pleasure, even if his genre—the 'merveilleux'—was deplorable (p. 155). The *Encyclopédie* article 'ballet' notes that when opera was established in France by Quinault, the basis of the 'grand ballet' was preserved, but its form changed: Quinault imagined a mixed genre of which recitatives formed the greatest part. (After 1681, when King and Court performed Quinault's and Lully's *Triomphe de l'amour* at Saint-Germain, grand ballet was no longer separately performed except in Jesuit colleges.)

The close associations between the Court and dance were demonstrated by Louis XIV's founding in 1661 of an academy for the latter art. The members, who numbered thirteen, either were or had been excellent dancers themselves. The academy of music was set up eight years later, and in 1671 Lully became its director. Compan notes the status enjoyed by dancing in seventeenth-century France when he observes in his article 'opéra' that several edicts of the Council in 1669 decreed that no gentleman or noble should forfeit his rank through becoming a dancer. The letters-patent of the Académie de danse referred to this art as being one of the most 'honnête' in which man could engage, necessary to form the body and give it a disposition for all sorts of exercise, such as using weapons. The emphasis on its utility is found elsewhere. Ménestrier mentions Plato's view that dance moderates the passions of fear, melancholy, anger, and joy, and is thus socially desirable: it lessens fear and melancholy by making the body more supple and easier to treat, and tempers anger and joy by soothing their impulses through regular movement. Furthermore, according to Ménestrier, it may directly help men in exercising their profession, for 'Les actions des orateurs, les cérémonies

---

[10] Noverre, *Lettres sur la danse et les arts imitateurs* (Paris, 1950), p. 122.
[11] [Compan,] *Dictionnaire de danse* (Paris, 1787), p. xi.

publiques, et l'exercice des armes demandent cette application pour acquérir cette souplesse de corps, cette adresse de mouvements et cette éloquence extérieure que Cicéron et Quintilien ont si fort recommandées' (pp. 33–4).

An undated *Mémoire sur les danses chinoises, d'après une tradition manuscrite de quelques ouvrages de Confucius*, despite the alleged origin of its observations, sets itself in the Platonic tradition on which Ménestrier draws, for it notes that 'Les anciens . . . virent [que la danse] perfectionnait l'âme, en mettant de la proportion, de la mesure et de l'accord dans ses mouvements.'[12] It mentions Plato's division of dance according to its various uses: military (to make the body strong for war), domestic (having agreeable relaxation as its object), and 'moyenne' (for the purpose of conducting expiations and sacrifices) (ibid.). The handbook *Le Maître à danser, qui enseigne la manière de faire tous les différent pas de danse dans toute la régularité de l'art, et de conduire les bras à chaque pas*, published in 1726 by Pierre Rameau (dancing-master to the pages of the Queen of Spain), notes that although dance was invented for pleasure—and thus, by implication is not natural—it contributes to the public weal. Rameau repeats earlier observations on the usefulness of exercising the body which it involves, and adds that it plays its part in the magnificent public fêtes 'qui font les délices des peuples.'[13] It is instrumental, furthermore, in drawing foreigners to Paris, with attendant advantages both to the public purse and to the stock of Frenchmen in the world (p. ix). There is no court in Europe which does not have its French dancing-master, and the art is one in which the nation reigns supreme (ibid.). A treatise by another dancing-master, Malpied, apparently written in 1789, suggests a further advantage. In an enlightened age, and under a 'gouvernement poli', there is need for bodily graces, 'puisqu'elles sont à la société ce que la tolérance et les mœurs douces sont parmi les nations civilisées'.[14] Malpied asserts that dance belongs to all walks of life and all social ranks, and that this universal appeal makes it natural that it should form the basis of education, 'puisqu'[elle] est à l'homme d'une utilité première, et que selon les lois d'une sage institution on doit autant s'occuper des progrès de l'accroissement du corps que des

[12] *Mémoire sur les danses chinoises, d'après une tradition manuscrite de quelques ouvrages de Confucius* (n.p., n.d.), p. 246.
[13] Sieur Rameau, *Le Maître à danser, qui enseigne la manière de faire tous les différents pas de danse dans toute la régularité de l'art, et de conduire les bras à chaque pas* (Paris, 1726), p. iv.
[14] Malpied (maître de danse), *Traité sur l'art de la danse*, 2nd edition (Paris, n.d.), Avis de l'éditeur.

connaissances morales' (ibid.). But many earlier commentators had emphasized its function in elevated society particularly, and this courtly tradition continues to be felt throughout most of the eighteenth century. The introduction to Rameau's book announces that the work is directed not just at the young, but also at those 'personnes honnêtes et polies' who desire rules 'pour bien marcher, saluer et faire les révérences convenables dans toutes sortes de compagnie'. It is through dance, Rameau declares, that 'nous nous comportons dans le monde avec cette bonne grâce et cet air qui fait briller notre nation' (p. 2). Clearly, his intended readership is a socially privileged one.

The book on choreography by Guillemin, *Chorégraphie, ou l'art de décrire la danse* (1784), states how people's need for pleasure gave rise to the art of dance, described here as belonging to the 'amusements honnêtes' and 'plaisirs sinon nécessaires, du moins utiles, en ce qu'ils donnent de la sensibilité à l'âme, et de la souplesse au corps'.[15] But the social dances which Guillemin proceeds to describe—the allemande, the minuet, and their like—belong to the milieu of the leisured classes. Compan, too, writes that dance 'caractérise seule une belle éducation' (p. viii), by which a socially refined education is obviously meant. Equally, the attendance at public balls, such as the Regent established at the Opéra, was predominantly of well-to-do citizens. The *Encyclopédie méthodique* notes in its entry 'bal' that the inauguration of such balls, which should have favoured the art of dance, was in fact inimical to it. The Opéra's efforts put an end to private dances, and their rivals at the Comédie-Française were unsuccessful in promoting their own version. The Opéra thus had no competition, and according to the *Encyclopédie méthodique* the consequence was that social dancing declined. Nevertheless, Baron regards the institution of the Opéra itself as having led to a broadening of the audience for serious dance (as opposed to the antics of tumblers and pantomime artistes at the fairs), and, correspondingly, to an extension of the social range of its practitioners. Before the creation of the Opéra, such dance was entirely reserved for the Court (p. 215). In 1726 Rameau remarks that, having described in *Le Maître à danser* 'les principaux pas des danses de ville' (p. 269), he will prepare a work on ballet movements 'tant sérieux que comiques', with a short introduction on ballet composition included, 'afin que cette noble jeunesse ne se trouve pas

---

[15] Guillemin (maître de danse), *Chorégraphie, ou L'Art de décrire la danse* (Paris, 1784), p. vi.

embarrassée lorsqu'elle sera obligée de paraître dans les ballets du Roi' (p. 270).

Towards the end of the century a more 'popular' theatre than the Opéra ventured to stage opera and ballet, at the Porte Saint-Martin, and high society flocked to the boulevard playhouse (Baron, pp. 287–90). But the critic Geoffroy observes a distinct difference between the ballets at this theatre and those of the Opéra. The contrast between the performance of Gardel's *Paul et Virginie* at the latter, and of Aumet's parody *Les Deux Créoles* at the Porte Saint-Martin, was that between ballet proper and pantomime (*Cours de littérature dramatique*, VI. 134). The production of *Les Deux Créoles* demonstrated Geoffroy's belief that

> Il ne faut ni talent ni mérite pour gesticuler au hasard, pour faire toutes les singeries de la joie, de la douleur, de la crainte, de l'amour, etc. La pantomime ne serait pas un art s'il n'était question que de rouler les yeux, remuer les bras, frapper du pied, se défigurer par des contorsions, des grimaces; il faudrait en ce genre céder la palme aux singes, qui sont de très grands pantomimes. (Ibid.)

Genius, on the other hand, consists in choosing among the different signs of passion those which suit the age and personality of the character; and these signs must be judiciously placed, not multiplied. If characters were really moved, he writes, they would be far more sober in their demonstrations, and would not exhaust themselves with convulsive movements. Nothing betrays a cold actor so much as exaggeration (ibid.). Lucian's treatise on dance, on the other hand, shows what study was required of ancient practitioners of the art, whereas 'on ne conçoit pas comment des comédiens de boulevard, souvent choisis au hasard et à la hâte, et n'ayant pour tout acquis que la routine du métier, pourraient exceller dans un genre aussi difficile' (p. 135). At the Théâtre Saint-Martin, all is fracas, 'gros traits qui marquent de loin' and 'un fatras de pathétique qui ébranle la multitude' (ibid.). The unsubtlety of performance reminds Geoffroy of his critical duty: 'quand je vois des partisans maladroits vouloir l'ériger en rival de l'Opéra, et . . . lui donner même la supériorité, je crois lui rendre un service essentiel en lui rappelant son origine, en l'avertissant du ridicule et du danger d'une pareille concurrence' (ibid.). Finally, for Geoffroy there can be no comparison between the leading theatres and their second-rate rivals:

Les théâtres secondaires ne doivent point s'écarter de leur genre; il ne leur appartient pas de rivaliser avec les premiers théâtres; et rien ne serait plus funeste pour des tréteaux que d'être jugés par comparaison avec les scènes régulières. Qui voudrait voir les ballets de la Porte Saint-Martin si on leur opposait ceux de l'Opéra?

Many contemporaries echo the sentiment of Compan, who writes in the 'opéra' entry of his dictionary that dance has always been the most brilliant part of the Opéra's own spectacles, however great the merits of composers like Lully, Campra, and Destouches, and of librettists like Quinault, La Motte, and Danchet. Earlier in the century Rémond de Saint-Mard regretted its ubiquity in opera, observing that it had once been admitted only as a kind of illustration, and to form a part of the whole. It is now the principal element, and

étouff[e] les autres parties dont nous avons beaucoup plus affaire . . . c'est là un abus de l'Opéra auquel il serait le plus nécessaire à remédier. Le trop d'extension qu'on donne à la danse, le frivole qu'on y joint, fait languir l'action, au lieu de la soutenir, la gâte, et il est ridicule qu'on ne songe pas que l'Opéra étant une tragédie, c'est-à-dire un ouvrage fait pour toucher, tout ce qui s'écarte de ce but est déplacé, et produit en conséquent un effet désagréable. (p. 56)

But Favart, writing in April 1763 to Count Durazzo, declares that his search at the Opéra for a 'première danseuse' has met with total failure, and that the ballet-master there has offered him no hope for the future: 'Le talent manque aujourd'hui partout, et notre Opéra, loin d'avoir des sujets surabondants, est fort embarrassé de faire des recrues pour lui-même' (II. 85).

> Il faut se rendre à ce palais magique [the Opéra]
> Où les beaux vers, la danse, la musique,
> L'art de tromper les yeux par les couleurs,
> L'art plus heureux de séduire les cœurs,
> De cent plaisirs font un plaisir unique.[16]

The fact that opera is a many-faceted art is taken by some commentators as an indication that it should not be judged by the same standards as apply to other spectacles. Quatremère de Quincy, for instance, writes that comparison between the stage action of opera and that of non-lyrical drama is inappropriate: 'Vous vous plaignez de ce que les

---

[16] Voltaire, *Le Mondain*, in *Œuvres complètes*, ed. L. Moland, 52 vols. (Paris, 1877–85), X. 86–7.

acteurs entrent, sortent et reviennent presque sans motif, c'est que vous jugez toujours les convenances musicales par les règles des convenances dramatiques; vous êtes toujours à la comédie, et jamais à l'opéra' (op. cit., p. 33). The truth, he argues, is that each art addresses the soul in its different way, and affects it by the means proper to that art. Where a number of expressive means are employed, as in opera, there is an overall weakening of the individual parts. The 'principe de l'unité de l'âme' means that the soul 'ne peut jouir de deux plaisirs égaux à la fois, ni supporter ensemble deux passions également fortes; que toute impression qui se divise s'atténue, et qu'en général la réunion de plusieurs sensations est la preuve ou de leur faiblesse, ou de la légèreté de l'âme qui les reçoit' (pp. 29–30). It is therefore unjust to complain that the *actio* in an opera is inferior to that of silent ballet. But to set against this truth, Quatremère continues, is the fact that the illusions conveyed by all the arts are more or less restricted, so that no one may rationally accuse sculpture of being without colour, or painting of lacking a third dimension, or pantomime of being silent. Yet

> par une compensation assez extraordinaire, les moyens qu'ils ont de les [illusions] produire sont d'autant plus actifs, qu'ils semblent plus invraisemblables. Malheur, au reste, à qui exigerait de chaque art des illusions complètes! Si l'illusion arrivait jusqu'à être entière, elle cesserait d'être agréable . . . Gardons-nous de chercher à détruire les invraisemblances qui tiennent à leur essence, et d'exiger d'eux une espèce de vérité qui serait plus fallacieuse encore que les contre-vérités qu'on leur reproche. (p. 31)

Certain 'unnatural' conventions are recognized by critics to be appropriate to the art of dance. In its pure form, the wordlessness of ballet means that a certain enlargement of *geste* is requisite for the 'meaning' of the work to be conveyed.[17] The article 'ballet' of the *Encyclopédie méthodique* echoes the sentiment expressed by Horace's *Ars poetica* that certain forms of an art must be enlarged in execution because of the distance from which they are perceived by spectators, and applies it to dance: 'Notre art est assujetti . . . aux règles de la perspective; les petits détails se perdent dans l'éloignement. Il faut dans les tableaux de la danse des traits marqués, de grandes parties, des caractères vigoureux, des masses hardies, des oppositions et des

---

[17] See Roland Virolle, 'Noverre, Garrick, Diderot: pantomime et littérature', *Motifs et figures* (Paris, 1974), p. 209, on the presence of such exaggeration in Noverre's ballets.

contrastes aussi frappants qu'artistement ménagés.' (This observation recalls Geoffroy's remarks on the 'crudeness' of ballet at the Porte Saint-Martin, but in a different spirit.) Because it is also a vocal and verbal art, opera is theoretically less in need than pure dance of such enlargement of gesture. Nougaret is something of an exception in arguing, as he does in his *Traité du geste*, that opera singers should use gesture with insistence because their words are not easily heard (in *La Littérature renversée*, p. 53). A slightly earlier work, the anonymous *Essai sur l'opéra* of 1772, concludes that gesture should be sparing in opera, and compares the latter with non-lyrical drama. The author remarks that the words of opera, whether in recitative or aria, are delivered less quickly than those of ordinary declamation, and the speed of gesture must correspondingly be reduced. On the other hand, he writes, gesture should be more marked than in conventional acting. The importance of action in operatic delivery is here seen as paramount,[18] and the author notes that 'Le théâtre de l'Opéra devrait être une bonne école pour le geste. Car ceux qui sont chargés de la partie de la danse, qui n'est autre chose que l'art du geste, le possèdent aujourd'hui dans un haut degré de perfection' (p. 131). Equally, he continues, those ambitious of success on the operatic stage should assiduously observe the actors du Théâtre-Français, for lessons not only in the art of gesture, but in every aspect of declamation. Such talents are as indispensable at the Opéra as at any other theatre. The interest of an opera cannot derive solely from the words or the music: 'Il faut . . . que l'action soit représentée, et le succès de cette représentation dépend du talent des acteurs' (ibid.). Whether these ideal prescriptions found practical realization must be open to doubt, however. Certainly, various writers complain that the Opéra's dancers failed to develop their talents as the performers of non-lyrical drama had done.[19] The article 'courante' of the *Encyclopédie méthodique* declares that most dancers are inadequately educated, and contrasts them in this respect with actors. The latter are inspired by the nature of their profession to undertake 'un genre d'étude propre à donner, avec l'usage du monde et le ton de la bonne compagnie, l'envie de s'instruire et d'étendre leurs connaissances au-delà des bornes du théâtre'.[20] Actors are widely read in poetry, drama, and history, the

---

[18] *Essai sur l'opéra* (Paris, 1772), p. 132.
[19] *Idées sur l'opéra* (Paris, 1764), p. 8.
[20] Framéry writes in *De l'organisation des spectacles de Paris*, p. 30, that 'la qualité la plus essentielle à un chanteur de l'Opéra, c'est d'être bon acteur.' See Dene Barnett,

author adds; and for dance to become perfect, its practitioners must similarly cultivate their minds. Mere physical prowess is insufficient. The dancing-master should provide his pupils with an example in this respect: according to Noverre's *Lettres sur la danse*, which here expresses an opinion elaborated in Lucian's treatise on dance, he needs the skills of the painter in mixing colours for contrast, for grouping figures and clothing them appropriately, and for giving them character and expression (pp. 69–70). Baron looks back to the masters of dancing in antiquity, who not only understood the theory of pantomime and executed it to perfection, but were knowledgeable in the fields of music, geometry (in order to plot movements), moral philosophy and rhetoric (so that they could depict mores and excite passions), painting and sculpture, and history and myth. Baron writes that he too has drawn all these requirements from Lucian (p. 121).

At the same time, the purely physical realization of dance was the concern of many handbooks on the art. Rameau's work is essentially a description of the way to execute the main society dances, and is accompanied by a number of engraved figures (by Rameau himself) illustrating the different movements. Baron mentions Noverre's proposal to record the physical motions of dance in such a way that they might be passed on to later ages. This would allow the usual fate of dancers and teachers to be averted; for they 'ne laissent jamais après qu'ils ont abandonné le théâtre qu'un souvenir confus des talents qui faisaient l'admiration de leur siècle' (op. cit., p. 195). Noverre's *Lettres* does indeed discuss this matter, and suggests that an artist might sketch groups of dancers in performance, and that these could then be engraved and thus multiplied for use in teaching. An interest in the graphic recording of dance was evidence much earlier than this, however. Feuillet's *Chorégraphie ou l'art de décrire la danse par caractères, figures et signes démonstratifs* (1700) popularized a system which Pierre Beauchamps had developed and circulated in manuscript form twenty-five years earlier. Baron notes that Beauchamps was declared the inventor of the art of choreography by an 'arrêt' of the parlement; but he also reports Guillemin's opinion that deciphering Feuillet's system of notation was a very laborious business (Baron, pp. 198–201). The *Encyclopédie* article 'chorégraphie' defines its

'Die Schauspielkunst in der Oper des 18. Jahrhunderts', *Hamburger Jahrbuch für Musikwissenschaft*, 3 (1978), 291.

subject as the 'art de décrire la danse comme le chant, à l'aide de caractères et de figures démonstratives'. The ancients allegedly knew of no such art, or at least failed to mention it, and Furetière's dictionary was the first to record the word (ibid.). Malpied's study of 1789 illustrates dance movements not by representing human figures, as do Rameau and Guillaume, but by a system of lines and symbols. Baron, finally, notes that Noverre's plans never reached fruition, and asserts that a choreographic academy should have drawn steps and traced the lines followed by dancers, and a verbal stylist explained what geometrical plans cannot convey distinctly, analysing the sequence of steps, the position of the body, attitudes, and mute 'pantomime'. Then a competent draughtsman could have set on paper the main groups and most interesting situations of ballet.

Mlle Clairon remarked that dance was a rule-governed art, and might therefore be taught in a way that acting could not.[21] Ménestrier mentions, without going into details, the 'règles qu'Aristote, Platon, Plutarque et Lucien nous ont laissées pour la conduite de ces représentations' (op. cit., preface), and historical investigations such as Bourdelot's and Bonnet's *Histoire générale de la danse* and Cahusac's *La Danse ancienne et moderne* describe some of the points of ancient doctrine. Many writers insist that in dance, as in other arts, a slavish following of rules can never lead to greatness, and that sentiment is an essential quality in the performer (e.g. Baron, pp. 284–5); but for all that, it is recurrently stated that rules should never be spurned. Nearly all the commentators refer to Beauchamps's establishing of five rules for the basic positions of dance. 'Position' is itself described by Rameau as 'une juste proportion que l'on a trouvée d'éloigner ou de rapprocher les pieds dans une distance mesurée' (p. 4), and Baron reports Rameau's observation that all other steps derive from these fundamental attitudes. Rameau's own rules, for the reasons already mentioned, are much concerned with the execution of the social graces, and he offers precise instructions on the correct manner of bowing and curtseying, doffing and donning a hat, and so on. Furthermore, in the interests of polite social intercourse, 'il est essentiel de savoir se poser le corps dans une situation gracieuse', and the means of doing this are described (p. 2).

Rémond de Saint-Mard complains in his *Réflexions sur l'opéra* that there is a sad uniformity in French ballet, which ultimately produces

[21] See chapter 6.

boredom and weariness in the spectator. He implies that the conventional nature of French dances makes them incapable of imitating situations or telling a story. In Saint-Mard's words there is an anticipation of the revolution which Noverre was to effect with his 'action ballets':

> Nos danses sont presque toutes dessinées les unes comme les autres . . . je ne dis pas que nos danseurs devinssent tout à fait pantomimes, ce serait trop: mais y aurait-il du mal qu'ils le fussent un peu? Qui les empêche de mettre de la noblesse dans leurs airs de tête, de l'expression dans leurs mouvements, de varier les attitudes, et de n'être plus enfin comme des danseurs de carton, qu'on fait remuer par machines! (pp. 93–4)

Compan repeats this notion: action, he writes in his dictionary entry for the same title, is in dance the art of conveying human passions and feelings through movement, gesture, and facial expression. It is, he continues, nothing other than pantomime. Every gesture and attitude should have a different expression, and it is not enough for the dancer simply to know steps: these, as Saint-Mard suggested, should correspond with the action and the soul of dance. Where the mood is light-hearted, steps will be swift; where it is grave, they will be heavier and slower. Compan's brief entry for 'bras' notes that Beauchamps was probably the first to give rules for arm movements in dance. Compan's section on 'geste' is more extended, giving some exact instructions for expressive use of the arms, but concluding with reservations matching Saint-Mard's: 'tant qu'on ne variera pas davantage les mouvements des bras, ils n'auront jamais la force d'émouvoir et d'affecter.'

Facial expression receives separate treatment in most handbooks, as it does in manuals on acting. Noverre's influence with respect to the abandoning of masks is well known, but Gardel had preceded him in this matter. When in 1768 the dancer Vestris wore a mask in order to please some nobles, his public found the practice ridiculous. Geoffroy deems exaggeration in facial expression—grimacing or rolling the eyes—as offensive as other forms of physical excess (*Manuel dramatique*, p. 134). In his entry 'sensibilité', Compan harks back to the physiognomical tradition given wide currency by Lebrun in his lectures on the facial expression of passions. He writes that the dancer endowed with sensibility will have a multiplicity of such expressions, each conveying a different emotion. His eyes, brows, and mouth are the 'sûrs interprètes des sentiments de son cœur'. Compan finishes this entry with an observation about the universal nature of the dancer's expressive language which differs not at all from those already noted in connection with pantomime and *geste*: 'le siège de son âme est inconnu, mais dans ses pas, ses gestes, ses attitudes, ses manières, &c., elle parle à tous les yeux , et son langage n'est point obscur' (ibid.). Like other performing artistes, the dancer is required to remain 'in nature', while emphasizing some aspects of his action to compensate for the non-verbal character of his performance. The *Apologie du goût relativement à l'opéra* of 1754 refers critically to the art of bodily eloquence as evinced in Italian opera. The principal singer is the castrato, while 'les acteurs subalternes sont réduits au récitatif, et font des grimaces que les Italiens appellent des gestes.'[22]

Before examining Noverre's objections to the artificiality of the dance tradition he found in mid-eighteenth-century France, and his attempts to confer a different character on the art, it is appropriate to consider what other writers say about the desire for effect, at the expense of true expressiveness, evident in some of their contemporaries. Vestris is disapprovingly mentioned in this regard by the author of the *Idées sur l'opéra* (1764), although an entry in Bachaumont's *Mémoires secrets* for 12 December 1770 notes his excellence in the pantomime inserted into the opera *Ismène et Isménias*. Bachaumont writes that Vestris, wearing no mask, 'a étonné le public par l'énergie de son exécution, non seulement comme danseur, mais encore comme acteur. Il met dans son personnage tout le sublime qu'on y peut désirer. Les passions se peignent sur son visage avec une noblesse, une vérité, une diversité qu'on ne saurait rendre, et qui décèlent en lui un talent singulier pour la scène.' But the earlier work asks with reference to this dancer, 'Pourquoi veut-il sauter? Pourquoi tendre plutôt aux grâces minaudières d'une femme qu'aux grâces nobles et majestueuses dont il a vu le parfait modèle dans le grand Dupré?' (p. 18). Previously, the author had mingled praise with blame in his discussion of Sophie Arnould's performances in opera. She is right, he observes, to believe that recitative should be accompanied by bodily action, but should guard against the temptation to exaggerate: 'Puisse-t-elle ne jamais outrer et se souvenir qu'il n'est permis qu'au pantomime de charger son jeu, parce qu'il ne parle point' (pp. 7–8). Dupré's art answered to the author's own beliefs about how heroic ballet (as opposed to ballet 'de caractère') should be performed.

[22] *Apologie du goût français relativement à l'opéra* (n.p., 1754), p. 18.

Regularity was the cornerstone of his performance: graceful arm-movements and noble, majestic attitudes seemed to this great dancer preferable to 'ces élévations forcées auxquelles on ne parvient que par des élans de bras convulsifs, suivis ordinairement de déhanchements, de jets de jambes en lignes croches, qui brisent l'aplomb que tout corps debout doit conserver, quelque position qu'il prenne' (p. 16). The affectedness of his successors would have dismayed him, according to the anonymous author. Nothing was further from Dupré's own conception of noble ballet than for dancers to

> tendre les bras aux loges, figurer des palpitements lascifs de cœur par des gonflements et abattements successifs de poitrine, et affecter tous les lubriques contours de corps d'un suranné blondin de Cythère. [Il n'eût] pu voir ces prétendus poèmes en ballet grimacés par des groupes d'hommes et de femmes, grotesques de taille et de figure, dénués d'oreilles, dansant des bras et remuant les jambes en disloqués, ou en véritables marionnettes. (p. 17)

To dance in this way, according to the author, is to demean a noble art, and put it on a level with the antics of the tumbler.

The author of the *Lettre critique sur notre danse théâtrale* of 1771 takes a different view of the old 'noble' style, however, finding it wanting in expressiveness. Its deficiencies, he suggests in the spirit of Noverre, could be made good if such dance were to be combined with pantomime. 'Les pas de deux dans ce genre noble, pas qu'il serait si aisé de rendre intéressants par la pantomime, ne peignent presque jamais rien' (p. 18). The arm and leg movements of the dancers convey very little, while their physiognomies are blank; and

> pour leurs yeux, il n'y faut pas songer. Comme ces gens-là n'ont rien à exprimer, leur physionomie devenant inutile, on a eu grand soin de leur couvrer la figure avec un visage de plâtre enluminé. En général, la danse grave et noble n'exprimera jamais rien, si vous ne l'animez point par une action que vous saurez y joindre. (Ibid.)

The faults of the old style could not be more effectively defined. It does not deal with actions and stories; it has, in fact, no story. This is close to the criticism levelled by many writers at the institution of society dancing, namely that it described only itself. Diderot had made this objection in the third *Entretien sur 'Le Fils naturel'*, where he observed that dance was still awaiting the man of genius: 'elle est mauvaise partout, parce qu'on soupçonne à peine que c'est un genre d'imitation. La danse est à la pantomime comme la poésie est à la prose, ou plutôt comme la déclamation naturelle est au chant. C'est une pantomime mesurée' (*Œ*, p. 162). The minuet, allemande, sarabande, and their like represent nothing, and again the fault lies in the fact that theorists and practitioners of dance are unaware of its representational character. The *Lettre critique* reports Rousseau's opinion (stated in his *Dictionnaire de musique*) that all dances which depict nothing but themselves should be banished from the lyric stage, and its observations on the limitations of lyric dance echo Diderot's (p. 7). The anonymous author also refers dismissively to the category of dance which he calls 'simple'. This type is performed without props, and in his view is no better than 'saltation', and should be relegated to the noisy assembly of a carnival. Only 'composite' dance is suited to the theatre (ibid.). Further on the writer refers directly to Diderot's ideas. He has just returned from the Comédie-Française, he writes, where he attended the first public performance of Diderot's *Le Fils naturel*, and was chagrined to find it poorly received. But 'Les scènes sublimes de cette pièce intéressante avaient excité dans mon cœur un attendrissement et une mélancolie que je cherchais à prolonger' (p. 25). Consulting anew his copy of the *Entretiens*, he was astounded to see that Diderot had anticipated there all of his own ideas on theatrical dance (p. 26). Compan's later dictionary, under the heading 'ballet', makes the same observations as Diderot about imitation. Although society dance may represent only itself, Compan observes, theatrical dance should be the imitation of some other thing. The idea was not a new one, however. Towards the end of the seventeenth century Ménestrier had noted that ballets imitate in movements 'les actions des hommes, leurs affections et leurs mœurs, comme ils imitent les mouvements naturels des animaux, et ceux que reçoivent naturellement ou violemment tous les autres corps' (p. 153). This, he says, is why Plutarch describes ballet as mute poetry which expresses its theme through gesture and movement. Ménestrier's earlier *Traité de tournois, joustes, carrousels et autres spectacles publics* (1669) remarks that 'Les ballets sont des représentations harmoniques et cadencées des choses naturelles, et des actions humaines.'[23]

Noverre believed drama and the action ballet to be closely related to one another, and was influenced by Diderot's writings on the theatre and by the experience of seeing Garrick's pantomimic acting style (Virolle, p. 211). He regarded some of the 'classical' conventions

[23] Ménestrier, *Traité de tournois, joustes, carrousels et autres spectacles publics* (Lyon, 1669), p. 7.

governing drama as unsuited to dance. Following Ménestrier, for example, he denied that the three unities should be observed in ballet, although he declared the need for action ballets to have an exposition, 'nœud', and 'dénouement'. As Diderot intended for the bourgeois *drame*, he wanted ballet to be based on an observation of ordinary life, and enjoined the ballet-master to attend to the physical accompaniments to everyday actions:

> Que de tableaux diversifiés ne trouvera-t-il pas chez les artisans! Chacun d'eux a des attitudes différentes, relativement aux positions et aux mouvements que leurs travaux exigent. Cette allure, ce maintien, cette façon de se mouvoir, toujours analogue à leur métier et toujours plaisante, doit être saisie par le compositeur; elle est d'autant plus facile à imiter qu'elle est ineffaçable chez les gens de métier. (*Lettres sur la danse*, p. 73)

But the world offered to his attentive gaze is not of one social group alone. Much can be learnt from watching 'la multitude de ces oisifs agréables, de ces petits maîtres subalternes qui sont les singes et les copies chargées des ridicules de ceux à qui l'âge, le nom, ou la fortune semblent donner des privilèges de frivolité, d'inconséquence et de fatuité!' (ibid.) (The remark is a striking anticipation of what Diderot will later say of the arch-mimic Lui in *Le Neveu de Rameau*, who 'degrades' others through his pantomime, and teaches the world to know itself better.) A stronger contrast with the instructions given to practitioners of society dance, and with the social circumstances for which their training prepared them, could hardly be imagined. Like Diderot, Noverre commends a certain freedom from restriction when he declares his dislike for merely mechanical virtuosity in dance on the grounds that it smacks of the rule-book.[24]

Plato's assertion that the practice of dancing may moderate the passions will be remembered. Previously to Noverre, Dubos, who similarly saw the performed art of dance as an animated painting, had remarked on its ability to exteriorize as well as govern human passions by the rhythmic imitation of actions which they engendered. Cahusac's study of ancient and modern dance likewise anticipated Noverre's assertion that dance should depict emotion;[25] and later

---

[24] See Manfred Krüger, *J.-G. Noverre und das „Ballet d'action": Jean-Georges Noverre und sein Einfluss auf die Ballettgestaltung* (Emsdetten, 1963), p. 10.

[25] Cahusac, *La danse ancienne et moderne, ou Traité historique de la danse*, 3 vols. (The Hague, 1754), III. 168. Towards the end of Louis XIV's reign, Rebel invented a new balletic style, that of 'symphonies dansées', which foreshadowed the mimographic

writers continued to state this belief. The *Lettre critique* of 1771 calls theatrical dance 'l'art de rendre les diverses impressions de l'âme par des mouvements variés des différentes parties du corps' (p. 6), and complains of the performance of Vestris and Heinel that it 'laisse notre âme dans la même situation où elle l'avait trouvée' (p. 12). A remark of Compan's in his entry 'passion' recalls a famous work executed in a different artistic medium. The expression of passions through dance is clearest, Compan writes, when the dancers' heads present a profile to the observer; and the reader is reminded both of David's painting *Le Serment des Horaces* (1784) and that Noverre's ballet *Les Horaces et les Curiaces* was almost certainly one of David's sources for the work. Geoffroy recalls to his reader that the poet Horace compared the difficulty of dramatic art to that of rope-dancing, and continues that rope-dancing has no pretensions to being an art, because it cannot express anything: 'et toutes les fois que la danse d'opéra n'exprime rien, elle est fort au-dessous de la danse de corde; celle-ci du moins est un grand objet de curiosité, un exemple étonnant de ce que peuvent l'adresse et l'industrie humaines' (*Cours de littérature dramatique*, VI. 148).

Further links between Diderot and Noverre are evident in their views on the desirability of observing pictorial principles in drama and ballet respectively. Diderot's theory of tableaux is echoed by what Noverre says in the *Lettres* about tableaux vivants (p. 226), which, he writes, have the advantage over conventional painting that they can portray a sequence of events.[26] Agreeing that ballet shows successive states, Noverre remarks that it may justly be compared with pictural series such as Rubens's sequence for the Galerie du Luxembourg, showing Marie de Médicis and the birth of her son. (Noverre differs from Diderot, however, in admitting into the fabric of ballet the coup de théâtre, which Diderot's theory excludes from drama in favour of the tableau.) According to the *Encyclopédie méthodique*, Quinault constantly built pictorial moments into his opera libretti which have not been turned to advantage in performance: 'le

dance of Noverre. In *Les Caractères de la danse* (1715) Rebel described dance as an independently expressive art (as it is described in Le Roy's *Ode de la danse* of 1714), declaring that it presented that mimesis of emotions which other arts influenced by Cartesian philosophy had pursued, and for which Lebrun had given a model in the visual arts.

[26] Cahusac (op. cit., III. 135–6) remarks that 'La danse théâtrale a tous les moments successifs qu'elle veut peindre. Sa marche va de tableau en tableau, auxquels le mouvement donne la vie.' See also Krüger, p. 19.

feu, le pittoresque, la fertilité des beaux cartons de Raphaël' are everywhere present in his work (article 'courante'). Compan's entry 'situation' enjoins the dancer of genius to begin his portrayal of a great action on the lyric stage by extracting from it all the situations susceptible of yielding a tableau: 'il n'y a que ces parties qui doivent entrer dans son dessin; toutes les autres sont défectueuses ou inutiles.'

Other writers comment on the analogies between painting and dance, which were mentioned as early as 1754 by Cahusac. He noted that the more objects dance (like painting) embraces, the more opportunities it has to show fine proportions and to confer on its subjects the movement which gives them life. 'Voyez que de jolis Teniers naissent chaque jour sous la main légère de De Hesse [a celebrated dancer]' (III. 143). Recalling other critical observations on the composite ballet's superiority to the 'danse simple', Compan writes in his entry 'opéra' of the contrast between the latter and the action ballet, declaring that they are equivalent to the portrait and the history painting respectively. (Mercier's comparison in *Du théâtre* of drama which depicts masses, and the well-filled canvas, with drama which concentrates on individuals, and the portrait, comes to mind.) Baron mentions the excavations at Herculaneum and Pompeii which uncovered pictures showing various types of dance-movements, all full of grace and lightness (p. 78). Later on (p. 236) he remarks that Noverre changed the monotony of ballets in which characters followed one another like a flock of sheep into a varied type analogous to certain paintings. Compan's comparison is with a series of related canvases where the arrangement varies from one frame to the next, and in which each group and character must adopt positions and execute movements different from all the others.

A part of Noverre's pictorial theory certainly preceded any general influence which Diderot's writings on drama may have exercised on him, for Collé notes in his journal for 1754 that the former's ballet *Les Fêtes chinoises* pleased its audience not with 'les pas et les entrées', but with an exciting range of 'tableaux diversifiés et nouveaux'.[27] For Noverre, as for other writers on dance, the new balletic style called for compositional integration of the kind that the isolated virtuoso efforts he rejected made impossible. In this respect his ideas resemble those of Diderot, whose tableaux are intended to supplant the sudden intrusion, destroying any sense of continuity, of the coup de théâtre. Auto-

---

[27] Charles Collé, *Journaux et mémoires*, ed. Honoré Bonhomme, 3 vols. (Paris, 1868), I. 428. Diderot's theory of the dramatic tableau was not published until 1757.

matism, which Noverre stigmatizes in the dancer and Diderot in the actor (for he favoured the latter's spontaneity in performing his role as much as his obedient attention to the playwrights's prescription), is clearly disliked by later writers too. The predilection for 'sensible', pathos-arousing art which makes the author of the *Lettre critique sur notre danse théâtrale* so deplore the cold reception of *Le Fils naturel* in 1771 equally causes him to disapprove of mere artistry in dance. And Baron notes that the French are still slow to appreciate dancing 'qui exige une âme, et pour laquelle il ne suffit pas d'être un pantin bien organisé' (p. 126).

The need to avoid such monotony as, according to Baron, characterized the dancing of Javilliers, Dumoulin, and Dupré, with their 'répétition perpétuelle des mêmes gestes et des mêmes figures' (p. 231), dictated the employment of certain pictorial principles in the composition of action ballets. Diderot's *drame* theory included the instruction that contrast should obtain in the constructing of tableaux or entire scenes, and his promotion of the idea that simultaneous sets should be employed in performances of drama rested on the belief that the portrayal of different but related actions presented the spectator with a visual richness unknown in individual portrayals. In similar fashion Diderot, like other eighteenth-century theorists of drama, rejected the principle of symmetry in composition. Noverre's *Lettres* likewise discourages the symmetrical placement of dancers on stage, and advises the student to observe the example of great painters to discover the principles of effective composition (p. 90). Noverre's recommendation concerns not just the grouping of characters, but also the painterly distribution of colours: the strong primary colours should take the foreground, and the more muted ones the space behind. In the *Encyclopédie méthodique*'s article 'ballet' it is noted that in the performance of Noverre's *Fêtes chinoises* these principles had been ignored, to the detriment of the whole: 'le mauvais arrangement des couleurs et leur mélange choquant blessait les yeux; toutes les figures papillotaient et paraissaient confuses, quoique dessinées correctement . . . Les habits *tuèrent*, pour ainsi dire, l'ouvrage, parce qu'ils étaient dans les mêmes teintes.' Everything shone forth with equal brilliance, with no part being sacrificed to another, 'et cette égalité dans les objets privait le tableau de son effet, parce que rien n'était dans l'opposition' (ibid.). The regular theatre, the author says, may ignore these pictorial rules, but not the lyric stage (to which, as we have seen, serious dance in eighteenth-century France was largely confined). The nature of opera—often hieratic, slow in movement, and naturally presenting in its arias long moments of concentration on a single scene—demands that they be followed in this type of theatre, 'où la peinture peut déployer tous ses trésors; théâtre qui, souvent dénué d'action forte et privé d'intérêt vif, doit être riche en tableaux de tous les genres' (ibid.). An article contributed by Grimm to the *Encyclopédie*, 'Du poème lyrique', similarly associates the tableau with the occurrence of arias in opera. Grimm observes that

L'air, comme le plus puissant moyen du compositeur, doit être réservé aux grands tableaux, et aux moments sublimes du drame lyrique . . . Le secret des grands effets consiste moins dans la force des couleurs que dans l'art de leur dégradation, et les procédés d'un grand coloriste sont différents de ceux d'un habile teinturier. Une suite d'airs les plus expressifs et les plus variés, sans interruption et sans repos, lasserait bientot l'oreille la mieux exercée et la plus passionnée pour la musique. C'est le passage du récitatif à l'air, et de l'air au récitatif, qui produit les grands effets du drame lyrique.[28]

Recommendations such as the ones already quoted concerning the desirability of unity in ballet imply their authors' belief that dance may be free-standing and autonomous spectacle, not necessarily the accompaniment to a larger whole. But this potential autonomy was for a long time denied. Where dance was one element of a composite art-form like opera, the need to make it an integral part was emphasized. The *Encyclopédie* article 'danse théâtrale' refers to the fact that the ancient Greeks united dance with comedy and tragedy, but notes that it was merely an intermezzo, not closely linked with the action. The *Encyclopédie méthodique*'s article 'ballet' mentions that Quinault succeeded in binding dance together with the main action in his operas, though later on it observes that, while opera could be an ideal vehicle for dance, it has not yet proved to be so. Composers have not felt the need to integrate ballets with their principal subject, and authors have too readily regarded them as hors d'œuvre, or used them to fill the intervals between acts. Noverre affirms that even where this remains their function, they could still be more directly related to the whole than they are at present (*Lettres sur la danse*,

---

[28] The article 'air' in Rousseau's *Dictionnaire de musique* (Paris, 1768) notes that 'Les airs de nos opéra sont, pour ainsi dire, la toile ou le fond sur quoi se peignent les tableaux de la musique imitative; la mélodie est le dessin, l'harmonie est le coloris; tous les sentiments réfléchis du cœur humain sont les modèles que l'artiste imite; l'attention, l'intérêt, le charme de l'oreille et l'émotion du cœur sont la fin de ces imitations.'

pp. 119–20). Diderot's *Entretiens* numbers among the theatrical reforms to be introduced 'la danse à réduire sous la forme d'un véritable poème, à écrire et à séparer de tout autre art d'imitation' (p. 167).

Some commentators, however, while supporting the idea that dance should be dissociated from the performance of opera, did so because they believed that its presence weakened the force of the latter. In *La Nouvelle Héloïse* Rousseau writes about the intrusion of ballet into opera where, in the course of every act, progress is halted at the most interesting point by the performance of dance before the seated singers. He then attacks ballet in sweeping terms:

> Non contents d'introduire la danse comme partie essentielle de la scène lyrique, ils [Parisiens] se sont même efforcés d'en faire quelquefois le sujet principal, et ils ont des opéras appelés ballets qui remplissent si mal leur titre que la danse n'y est pas moins déplacée que dans tous les autres. La plupart de ces ballets forment autant de sujets séparés que d'actes, et ces sujets sont liés entre eux par de certaines relations métaphysiques dont le spectateur ne se douterait jamais si l'auteur n'avait soin de l'en avertir dans un prologue.[29]

In the article already mentioned, Grimm likewise expresses disapproval of the interruption of operas by dance, and objects that

> cette partie postiche est même devenue, en ces derniers temps, la principale du poème lyrique . . . et le succès d'un opéra dépend aujourd'hui, non pas précisément de la beauté des ballets, mais de l'habileté des danseurs qui l'exécutent . . . Il faut que l'action de ce poème soit dénuée d'intérêt et de chaleur, puisque nous pouvons souffrir qu'elle soit interrompue et suspendue à tout instant par des menuets et des rigaudons; il faut que la monotonie du chant soit d'un ennui insupportable, puisque nous n'y tenons qu'autant qu'il est coupé, dans chaque acte, par un divertissement.

In 1776 the *Mémoires secrets* greets with disapproval Noverre's efforts to accustom the public to ballets independent of any supporting musical spectacle. It was not for this that the Opéra administration sought his (expensive) participation in their enterprise. The author notes that during this year people have been generally critical of Noverre's attempt in the ballet *La Générosité d'Alexandre* to 'faire le principal du spectacle de ce qui ne devrait faire que l'accessoire'. Noverre should turn his attention to perfecting the art of choreography, 'c'est-à-dire en trouvant l'art d'insérer dans nos opéra et autres ouvrages lyriques des danses analogues au sujet, faisant partie de

[29] Rousseau, *La Nouvelle Héloïse*, ed. René Pomeau (Paris, 1960), pp. 266–7.

l'action, en soutenant, en augmentant l'intérêt, et complétant la magie de ce délicieux spectacle' (IX. 237). The *Mémoires secrets* returns to the attack the following year, discussing Noverre's balletic transformation in *Les Horaces et les Curiaces* of Corneille's *Horace*:

> On doute que ce spectacle isolé ait le succès que s'en promet le Sieur Noverre, et s'il ne réussit pas cette fois on regarde sa mission comme manquée. Il est fâcheux qu'il ne veuille pas s'astreindre à composer des morceaux de chorégraphie adaptés à une action théâtrale, qui l'accompagnent, lui soient subordonnés et, sans la refroidir ou l'étouffer, la soutiennent, au contraire, et la réchauffent dans les entr'actes accordés à la danse. (X. 15)

But a few years earlier Bachaumont had seemed to approve the production of ballets detached from the supporting structure of opera. In his notice of 12 December 1770 on the same opera in which he found Vestris so impressive, he observes that the dance in which he shone was the creation of Noverre, 'l'homme qui ait le plus de génie en ce genre; mais il serait à souhaiter que cette pantomime s'exécutât seule, et ne fût pas réunie à un opéra, où d'accessoire elle devient l'objet principal par sa trop grande étendue, et la perfection de l'action.' Clearly, though, his desire for the separation of ballet here stems from his belief that its inclusion damages the opera. Later in the century, there remain doubts about the wisdom of trying to marry dance to some larger spectacle. The author of the *Entretiens sur l'état actuel de l'Opéra de Paris* (1779) comments sternly that 'De tout ce qui constitue ce qu'on appelle spectacle, la danse est certainement ce qu'il y a de plus postiche et qui se lie le plus difficilement au sujet'[30]—an observation firmly in the tradition of Rousseau's criticisms. In 1789 Quatremère de Quincy notes that the French have still not succeeded in providing intrinsic interest in the ballets which adorn their spectacles. In this connection he contrasts his countrymen with the Italians, for whom dance without dramatic interest is inconceivable (p. 36). The majority of ballets which grace spectacles in France are 'des espèces de lieux communs, dénués d'intérêt et surtout d'intérêt dramatique . . . enfin c'est la danse qu'on applaudit, et fort peu l'intérêt d'une action presque nulle' (p. 37). The virtuoso balletic performance, it seems, is still common, and 'On danse à Paris pour danser, comme en Italie l'on chante pour chanter' (p. 38).

Yet many writers are convinced that action ballets, like certain dramas, have the potential to convey great subjects: not the elevated

---

[30] *Entretiens sur l'état actuel de l'Opéra de Paris* (Amsterdam, 1779), p. 128.

5 David, *Les Sabines*. Photo: Bulloz

6 Carle Vanloo, *Mademoiselle Clairon en Médée*. Photo: Staatliche Schlösser und Gärten, Potsdam–Sanssouci

7 Carle Vanloo, *Médée et Jason*, sketch. Photo: M.-L. Pérony

8 Watteau, *The Mountebank*. Photo: Ashmolean Museum

states of the old 'ballet noble', a creation of and for the Court, but the mores of the people. Compan's entry 'mœurs' notes that in ancient Greece dance was a figural imitation of existence, although he provides no details about the way in which such imitation was effected. Guillaume's *Almanach dansant* of 1770 makes the same broad claims, again with no real substantiation. 'La représentation des ballets,' Guillaume writes, 'est une manière d'instruction pour la règle des mœurs, qui imite par les mouvements de la danse toutes les actions humaines.'[31] But the intention voiced by Cahusac and then by Noverre, to make dance representative of a reality external to itself, is clearly present throughout the second half of the eighteenth century, and provides a link with the intention of dramatists and actors to make stage action expressive of social circumstances. As in dramatic theory, the emphasis on the power of bodily movement to affect the audience's sensibility and to stimulate its thoughts is strong. But it seems that Noverre, like Diderot, failed to present in many of his works a truly convincing portrayal of ordinary lives. Indeed, his ballet *L'Amour corsaire*, which borrows from *Le Fils naturel* its main characters and an element in its plot, is even further removed from the everyday world than Diderot's *drame*. The true affinity of its story, which tells of shipwreck, bloody sacrifice, and providential escape, is with the romanesque literature of the previous century.

While revolutionary in theory, then, Noverre's works were in practice often as wedded to tradition as Diderot's *drames*. But the fact that this indebtedness was principally to a literary model is of some importance. However unpromising the narrative material contained in Noverre's ballets may have been, they were innovative in their very effort to describe action. The story-line was often implausible and far from the everyday world he encouraged dancers to observe, but its mere presence was enough to remove the ballets from the non-dramatic realms which dance had previously inhabited. Furthermore, despite contemporary opposition to his efforts, Noverre succeeded in breaking the mould of performing style in ballet, and did so more decisively than Diderot was able to do in the regular theatre. Possibly this was because the necessity of learning the movements of dance was generally accepted, whereas the desirability of acquiring technique through precept in conventional acting was not, so that a teacher's influence could more easily make itself felt in the former

---

[31] Guillaume, *Almanach dansant* (Paris, 1770), p. 8.

than in the latter art. Whatever the truth of the matter, Noverre's legacy to the dancing world was of balletic action in which the mute 'jeu' of the performer was eloquent as it had not been before.

CHAPTER SIX

Rules and Correspondences

Despite the scepticism which many eighteenth-century commentators felt about the possibility of training actors, attempts were made over the period to establish institutions for this purpose, and to settle the principles on which the actor's art rested. It was acknowledged that techniques for performing comedy differed from those of tragedy, but the Comédie-Française, at least, insisted that aspirants should distinguish themselves in both genres before being accepted as members of the company. There was never universal agreement among actors that formal training was necessary, but the number of attempts to create acting academies in the second half of the eighteenth century suggests a fairly wide acceptance of the notion. The desire to see acting granted its own academy was, I have suggested, part of a more general conviction that the profession deserved parity of esteem with existing academic arts. The belief, to which many subscribed, that acting could be partially if not wholly taught according to a system of rules led to the drawing of comparisons with the various liberal arts whose own procedures were rule-governed. Such comparisons, especially with the arts of oratory and painting, often involved consideration of the importance of gesture and attitude in the work of the performer or artist.

In 1738 Luigi Riccoboni expressed surprise that no nation had thought of setting up drama schools (pp. 43-4), and suggested that an orator of advanced years should be appointed to a chair in declamation. His lessons, Riccoboni wrote, would be as useful to society as those of the finest educational institutions (p. 44). But this was not a sentiment universally shared at the time by actors themselves. Both Lekain and, later, Molé, although each was to be associated with efforts to create training schools for actors, claimed that experience of acting in the provinces had once been an excellent and sufficient preparation for the man or woman ambitious to perform on the Paris stage. According to Lekain, provincial theatres were a 'milice réelle, de laquelle on pouvait tirer les meilleurs sujets pour compléter la troupe du Roi'.[1] Later in the century Mlle Raucourt of the Comédie-Française recommended that students of acting 'aillent se former dans les grandes villes où la fureur du spectacle a rendu les habitants très difficiles et très éclairés sur tout ce qui tient au talent dramatique.'[2] In her memoirs Mlle Clairon denies the need for acting schools. Dancing and singing can and should be learnt, she declares (p. 269, p. 302), but the same is not true of acting: 'je ne connais point de règle pour apprendre à penser, à sentir; la nature seule peut donner ces moyens que l'étude, des avis et le temps développent' (p. 303). Critics of her style, however, claimed that it seemed studied, lacking spontaneity and the instinct for verbal and gestural eloquence. In *Le Neveu de Rameau* the Nephew remarks of the chess-player de Bissy, 'Celui-là est en joueur d'échecs ce que Mlle Clairon est en acteur. Ils savent de ces jeux, l'un et l'autre, tout ce qu'on en peut apprendre' (p. 7). But in the *Paradoxe sur le comédien* Diderot's 'porte-parole' declares that Mlle Clairon's rational approach to acting is the key to her superiority. He describes her art of performance as consummately skilful, and says that its perfection resides in her intellectual mastery of a part and emotional detachment from it. At the start of the nineteenth century Austin wrote that rules, in acting as elsewhere, were never for the guidance of those who had genius, but for those who lacked it. Clearly, the unwillingness of some actors and observers to allow that rules were necessary to successful performance was related to the notion that an art governed by genius alone was greater than one directed by reason. I shall return to this point shortly.

Lekain's beliefs were evidently changed by what he perceived to be the decline in serious acting brought about by the popularity of the Opéra-Comique from 1751 onward (Pierre, *Anciennes Écoles*, p. 6). He held that the easy success open to performers in this genre deflected the young from a serious study of stagecraft, and meant that there might soon be an insufficiently large reserve from which the Comédie-Française could draw its own actors. Lekain's concern with maintaining standards of performance led him in 1756 to submit to the Premiers Gentilshommes a memorandum 'tendant à constater la nécessité d'établir une école royale, pour y faire des élèves qui puis-

---

[1] See Théodore Lassabathie, *Histoire du Conservatoire impérial de musique et de déclamation* (Paris, 1860), p. 3.
[2] See Constant Pierre, *Le Conservatoire national de musique et de déclamation* (Paris, 1900); also Étienne *et al.*, *Almanach général*, I. 285.

sent exercer l'art de la déclamation dans le tragique, et s'instruire des moyens qui forment le bon acteur comique.' (*Mémoires de Lekain*, p. 174 ff.) Lekain expresses the same desire as other commentators to see acting granted the status that was conferred by Louis XIV on dance and music when he established academies for those arts (p. 177), and blames the neglect suffered by acting on the indifference of the Premiers Gentilhommes. Whereas they ought to be concerned with maintaining the glory of the French theatre, they have failed to show actors the consideration due to them, and since the death of Molière and the retirement of Baron have forbidden them access to the Court (p. 178). Lekain's submission failed to bring about the establishment of an academy in 1756, however. He tried again in 1759, presenting a detailed scheme for the proposed institution and once more stressing the unwelcome influence of the Opéra-Comique on standards of serious acting. Pupils who engaged to perform in this type of entertainment, he wrote, would be stripped of their rights and status (pp. 410–11). But for a second time Lekain's proposal failed to win official approval.

Although he himself had learnt his profession before the establishing of academies for acting, Lekain possessed a degree of cultivation that was the admiration of his colleagues and successors. While emphasizing the primary importance to the actor of inborn talent, Mlle Clairon notes that Lekain 'avait fait d'excellentes études; il savait plusieurs langues, lisait beaucoup et jugeait bien' (p. 245). Talma confirms the truth of this statement. Mlle Clairon stressed the breadth of education needed by all who were ambitious of success in the art of acting. She was convinced of the necessity for actors to possess certain forms of academic knowledge, and wrote that they should be familiar with history, ancient myth, belles-lettres, languages, and geography (p. 278). The study of history and of the relics of antiquity was particularly recommended by Tournon, in his handbook on acting, for the performance of tragedy; comedy, for obvious reasons, was seen as requiring a different kind of awareness in the actor: 'L'histoire et les peintures antiques sont les monuments durables d'où l'on peut puiser la vérité du cothurne tragique; quant à celui de la comédie, il suffit de l'usage du monde.'[3] However, it does not appear that when academies for acting were set up, the kind of

---

[3] [Tournon,] *L'Art du comédien vu dans ses principes* (Amsterdam and Paris, 1782), p. 135.

education in history that painters of the Académie royale de peinture et de sculpture received (Locquin, p. 82 ff.) formed part of the actor's training. Dorfeuille writes that the actor need not be learned, but should have a knowledge relative to his art: for example, he should have an understanding of 'l'esprit, les usages, les mœurs, les conventions reçues dans les temps les plus reculés'. Only thus will he learn how to depict men not as he sees them now, but as they really were (I. 22-4).

Dorfeuille's handbook includes a fairly extensive list of instructions in the use of attitude, gesture, and facial expression. He explicitly distances himself from the view of Mlle Clairon that no rules useful to the actor's art exist, and clearly counts *actio* among those aspects of performance in which instruction is both feasible and desirable (I. 34). Some writers suggested that training of this kind had been known in the ancient world. In the *Réflexions critiques sur la poésie et sur la peinture*, for instance, Dubos mentioned some indications in Quintilian that the Greeks had had schools in which aspiring actors were taught the art of gesture (I. 525). On the other hand, the general lack of instructions concerning *actio* was recurrently alluded to by theorists. Thus Dubroca, whose treatise on declamation was published at the beginning of the nineteenth century, declares that few authors have attempted to detail the rules appropriate to a training in the art of 'geste', although, as we have seen, Engel had tried to do so in his *Ideen zu einer Mimik* of 1785-6. This book was translated into French in 1788-9, and the actor Molé is known to have admired it. Another German, Lessing, had earlier wanted to write a work on 'körperliche Beredsamkeit', or body language, but his project never advanced beyond a fragmentary state.[4] Dubroca saw an obstacle to the developing of such a system as Lessing and Engel envisaged in the infinite variety of gestural nuance. Dubos, too, had discussed this problem in the *Réflexions*. In Dubroca's view, separate rules would have to be provided for each individual's practices in bodily communication, and this probably explains why he himself rarely undertakes the detailed prescription of *actio* in his treatise. Where he does, it is normally only to refer the student to instances where 'geste', in the extended sense of bodily expression in general, is exemplified in a work of visual art.

---

[4] See Theodore Ziolkowski, 'Language and Mimetic Action in Lessing's *Miss Sara Sampson*', *The Germanic Review*, XL (1965), 261-76.

Mention of the visual arts brings to mind a recurrent observation by writers on acting in eighteenth-century France, namely that the actor's performance suffers in comparison with many other types of artistic production by virtue of its ephemeral nature. Similar observations were sometimes made about dance, as we have seen. In his *Chironomia* Austin bemoaned the fact that little apart from a portrait by Hogarth and one by Reynolds existed to recall something of Garrick's art, and observed that uncertain and fleeting traditions had in their frail keeping all the rest of this great man's action (p. 279). He declared, furthermore, that no picture could do more than mark a moment of performance, as in the famous portraits of Kemble and his sister Mrs Siddons. Some art or invention was needed, he concluded, to keep pace with the public speaker, and represent with fidelity his manner of delivery and gestures, accompanying his words in all their various transitions and mutual relations (p. 280). While sculpture, painting, literature, and music remain to delight and instruct future generations, the actor's performance does not. (Obviously, the case of musical performance, as opposed to musical composition, more closely resembled that of the actor in an age which lacked modern techniques of recording.) Some of the regrets voiced by eighteenth-century commentators referred to the verbal eloquence of the actor, which is not my concern here, but many too concerned his bodily eloquence, or *actio*. It seems likely that a number of the appeals made for the foundation of acting schools originated in an awareness that the absence of permanent models of past achievement is disadvantageous to the student of acting.

The desirability of explaining how the great performers of the past had tackled a role was acknowledged, but the problem of finding a means adequate to the task was acutely felt. Engel, conscious of the difficulty of conveying action with words, tried to introduce diagrammatic representations of movement, similar to some of the systems already developed by writers on choreography. Austin remarks that Engel's description of a particular Italian gesture filled an entire octavo page of his book (*Chironomia*, p. 277), and concludes that gesture is very easy to understand, but not to describe. Austin writes of the clarity of the notational scheme which he himself has developed, and observes that the gesture so laboriously rendered by Engel could be far more economically conveyed by his own method, where symbolic letters preserved patterns of delivery. He hoped to perfect his notational code in such a way that every action of the orator delivering a speech, or of the actor performing in a play, could be set down for posterity.

Certain pictorial means of reproducing the look of stage action were, of course, available to the eighteenth-century student, but they did not necessarily convey the general style of an actor's performance. The fact, much discussed in the eighteenth century, that painting is a non-temporal art-form means that its usefulness as a recorder of theatrical performance is inherently limited. As my earlier remarks on the tableau suggested, a picture may accurately depict a more or less motionless scene from a play, but its ability to convey action is restricted. Furthermore, the artist commonly selects the 'fruitful moment' for depiction, and this moment may be that of highest emotion or greatest tension for the performer, when his *actio* is enlarged or made in some other way to contrast with his habitual state in the play. I have already mentioned Carle Vanloo's portrait of Mlle Clairon in this connection (pp. 91–3). The modern practice of producing posed 'stills' as advertisements for dramatic productions is possibly analogous to an eighteenth-century painter's depiction of his dramatic subject: it may be that neither truly represents a moment in the play as an audience would actually see it.

It seems a reasonable assumption that the quick sketch or series of sketches of a point in the action of a play will render its spirit more accurately than a highly wrought canvas depicting a particular significant (or 'dramatic') instant. The work of Watteau and, later in the eighteenth century, Saint-Aubin exemplifies this type of response to spontaneous action, on the stage and in real life. Watteau's fondness for drawing vignettes of everyday activity often reflects his perception of their dramatic quality. The sketch of a mountebank in the Ashmolean Museum, for instance, reveals his ability to capture a scene of action 'sur le vif', and suggests the theatrical element in popular spectacles of the eighteenth century (*see Plate 8*). The scene is set in a fairground, and shows a stage which a quack doctor has momentarily taken over from actors to sell his wares. His attitude and gesture are those of a skilled performer, and the monkeys sitting on the rope that holds up the stage curtain seem to admire his eloquence as much as the audience does. A sketch on the verso shows a different scene from the same 'performance', with actors standing behind the mountebank awaiting their turn. Their costumes, and the stage curtain, belong to the same world as the charlatan's own histrionic activity, and incidentally reveal that even during the enforced absence of the Italian troupe from Paris Watteau found plentiful models for comic theatrical sketches at the fairground. The Ashmolean sketch probably dates from 1706 or 1707, before the recall of the *Commedia* actors to Paris. Watteau, who died in 1721, can have been directly influenced by the Italians only in the last years of his life, although his *Commedia* figures now seem a quintessential facet of his art. Watteau was an incomparable draughtsman, and his skills were universally acknowledged during his lifetime. His friend Gersaint, a picture-dealer, predicted that his would always be recognized as an exceptional talent. Yet the comte de Caylus claimed that Watteau's art had a weakness, and that it resided in the shortcomings of his drawing. According to Caylus, this prevented him from painting or composing anything heroic or allegorical. Watteau did, in fact, occasionally produce paintings based on mythology,[5] and an allegorical element may be discerned in many of his best-known canvases. But whether or not he regrets Watteau's omission to paint many works of a directly mythological nature, the modern observer is unlikely to concur with Caylus's judgement concerning Watteau's abilities as a draughtsman. His sketches have a vigour and an impulse to movement which even his greatest paintings lack: compared with the studies in crayon, pen, and brush, his finished oils are immobile. This is as true of the paintings showing actors, such as *Gilles* and *L'Amour au Théâtre Italien*, as of the 'fêtes galantes'.

An ability like Watteau's to present a 'snapshot' image of arrested movement is apparent too in the drawings and etchings of Gabriel de Saint-Aubin, who depicted scenes from the Opéra, Comédie-Française, Comédie-Italienne, and Opéra-Comique. The vitality of the artist's line and the mobility of his chiaroscuro, quite apart from his subjects, suggest movement to an impressive degree. In Saint-Aubin's energetic work there is an intense commitment to the present, and his approach helped to inaugurate some of the fundamental changes in vision which were to be evident in the lithography and photography of the nineteenth century.[6] Saint-Aubin set a high value on compositional spontaneity and the portrayal of the passing instant, and was probably attracted to the theatre for this reason. Like Watteau in the sketch of the mountebank, Saint-Aubin was able to suggest the theatrical element in life even outside the playhouse. His ability to

---

[5] See, for example, his *Jugement de Pâris* (Louvre).

[6] See the catalogue to the exhibition *Prints and Drawings by Gabriel de Saint-Aubin* (Baltimore, 1975), p. 15.

render the moving figure, one of his greatest strengths, is apparent in the 1760 *Spectacle des Tuileries*, which depicts the bustle of Parisian life as though the crowds were strolling players, and where even the men pulling a water-cart seem to be taking part in a ballet. Saint-Aubin's temperament seems to have made him impatient with the lengthier process of etching, and it is in his crayon, pen, and brush sketches that he most successfully calls an image into action, captures movement, and seizes an expressive gesture.

In his 1764 handbook on acting, d'Hannetaire did not consider using the skills of the artist in an attempt to capture the style of a performer, but suggested that verbal description of *actio* could be employed more helpfully than was generally the case. Yet his proposal that an actor's interpretation of his role be noted directly on the text of the play in question, and a library of such texts built up for future performers to consult, in no way answers the problem of describing action on stage in words, even though he suggests, tantalizingly, that his method has already been practised with success:

> Qui empêcherait par exemple de faire des notes ou commentaires sur chaque pièce qu'on joue, afin de recueillir et de conserver, par ce moyen, la meilleure façon dont chaque personnage aurait été rendu par les plus habiles comédiens? Ne nous a-t-on pas déjà transmis, d'une manière frappante, certains passages sublimes et lumineux de feu Baron et de quelques autres célèbres acteurs et actrices? (p. 27)

Since the records to which d'Hannetaire refers seem no longer to exist, there is no means of discovering how effective they were, or precisely in which manner the actor's interpretation was noted. The difficulty of expressing action with words was acutely felt by those playwrights who tried to prescribe the movements of a character by means of stage directions within the play itself. In the previous century Molière had written of this problem: 'il est assez difficile de bien exprimer sur le papier ce que les poètes appellent jeux de théâtre, qui sont de certains endroits où il faut que le corps et le visage jouent beaucoup, et qui dépendent plus du comédien que du poète, consistant presque toujours dans l'action.'[7] In his third *Discours sur le poème dramatique* Corneille had remarked that the dramatist should note in the margin of his play 'les menues actions qui ne méritent pas qu'il en charge ses vers, et qui leur ôteraient même quelque chose de leur

---

[7] *Épître dédicatoire à un ami* (preface to *Sganarelle*) in *Œuvres de Molière*, ed. Eugène Despois, 13 vols. (Paris, 1873–1912), II. 158–9.

dignité, s'il se ravalait à les exprimer'.[8] The ancients, Corneille admitted, had not adopted this procedure, but their plays were full of obscurities which it might have lessened. A later critic reacted testily to the expedient developed by Diderot and those he influenced of furnishing stage directions in the printed text of a play. In his review of a three-act verse drama by Fabre d'Églantine, *Le Collatéral*, which was first performed in 1789, the anonymous commentator observes:

Un . . . travers choquant dans le *Collatéral*, mais qu'on retrouve aussi, quoique moins marqué, dans beaucoup d'autres pièces imprimées, c'est cette affectation prétentieuse de tracer à chaque ligne le jeu et la pantomime de l'acteur . . . On peut mettre de pareilles notes sur le rôle d'un acteur; mais le lecteur en est impatienté, et n'y voit que la petite charlatanerie d'un auteur qui rejette dans la pantomime l'expression qu'il n'a su mettre dans le style. Il est permis quelquefois d'indiquer l'esprit général d'une scène, dans les occasions importantes où les acteurs peuvent se méprendre. Voltaire l'a fait quelquefois; mais il y a loin de ces précautions rares à cette bigarrure continuelle de petits avertissements italiques, dont on noircit toutes les pages d'un drame. (*Mercure de France*, May 1792, pp. 100–1)

But the writer's complaint is less about the possibility of expressing gesture and movement with words than about technical incompetence or laziness (as he sees it) in the playwright's construction of the play. Sophie's dismissive comment about writers of pantomime in *La Pantomimanie* comes to mind. Diderot's expedient, in any case, was resorted to less as a result of a belief about the unclarity of words where expression of movement is at issue than because he held that drama was, in general, too wordy and insufficiently concerned with bodily expression.

Besides, as the *Entretiens* make clear, he also favoured the actor's ignoring on certain occasions all the playwright's prescriptions, and following his own impulses. In the second of the *Entretiens sur 'Le Fils naturel'* Dorval declares that there are moments in a play which the dramatist should leave to the performer: in such instances the latter should have full control of the written scene, repeat some words, return to certain ideas, cut some out, and add others (p. 101). Voice, tone, gesture, and action belong to the actor alone, and it is they that strike the audience most powerfully and convey the strongest sense of dramatic energy (p. 102). The section on pantomine in *De la poésie dramatique*, in similar vein, observes that when this art has been

---

[8] Corneille, *Writings on the Theatre*, ed. H. T. Barnwell (Oxford, 1965), p. 70.

perfected, the playwright will be able to leave it unwritten: the actors themselves, by implication, will possess sufficient skill to shape it adequately. This, it is suggested by Diderot, explains why the drama of the ancients is largely devoid of instructions on *actio* (p. 277). And even where the playwright does detail pantomime, Diderot continues, he must accept the right of the actor to know better than himself, and thus to improvise. A man of genius should not be treated like a machine (p. 278). In the *Paradoxe sur le comédien*, on the other hand, Diderot describes the actor as a marvellous puppet, which suggests that some fifteen years after the writing of *De la poésie dramatique* his ideas about autonomy in performance had undergone a change. Diderot's new view is that the poet holds the puppet's strings and indicates, in every line of the play, the actions he should perform. The actor, he observes, is like a courtier, a professional player of roles who lacks the exquisite sensibility of an original spirit.

But Diderot and other eighteenth-century playwrights and critics leave untouched the problem of the word and its limited evocativeness as far as the recording of *actio* is concerned. There are, it is true, occasional indications of the graphic manner in which an actor played his part. Talma is known to have been inspired in his playing of Hamlet by the antique sculptured group *Poetus and Arria* (which, as we have seen, was recommended as a model by various eighteenth-century French writers on bodily eloquence); and the Roman statue of Sulla in the Villa Negroni seems to have suggested poses to him, to judge by the fact that an extant engraving after this statue is included among the portraits of Talma at the Bibliothèque nationale.[9] It is worth noting that the actor himself in turn inspired artists. Guérin's *Phèdre* and *Pyrrhus* were apparently based on moments from Talma's performances, and Stendhal noted that Steuben's *Serment des trois Suisses* in the 1824 Salon showed its three heroes imitating poses of the great actor's too (ibid., p. 173). In chapter 3 I discussed the occasional imitation on the eighteenth-century stage of scenes and attitudes from painting. But in all such cases, as with Talma's rendering of antique models, the indication is only valid for the moment of the 'still', and tells us very little about the general style of performance.

Diderot's observation that actors should retain the right to extemporise in their performance, as the members of the *Commedia dell'arte*

---

[9] See Georges Wildenstein, 'Talma et les peintres', *Gazette des beaux-arts*, 55 (1960), 171–2.

habitually did, is echoed by what Noverre says about prescription in the art of dance. Noverre disliked any suggestion of automatism in a performer's action. In this, incidentally, he was echoed by Talma, who allegedly remarked that the actor Larive's talent was a mechanical one, governed by clockwork and thus comparable with the workings of Vaucanson's famous automata. Talma's damning opinion is that Larive utterly lacked soul, the essential component of the true performer's art (Regnault-Warin, p. 34). Noverre spoke out against choreography in one of his letters on dance because he perceived its limiting function, just as Diderot inveighed against the corruptions of academic routine among students of painting. Since he believed that dance should express 'nature', Noverre disapproved those forms of it that announced their artificiality; and this dislike too he shared with Diderot, whose dismissive comment on the attitude of a figure in one of Carle Vanloo's paintings—'On dirait qu'elle a été arrangée par Marcel' (*Salons*, II. 63)—has already been noted.

It could, of course, be argued by critics that *all* forms of academic teaching—of acting, painting, dance, or whatever—fostered a reliance on rules that was likely to prove artistically regrettable. The student, according to this view, would grow accustomed to neglecting the direct observation of life that Diderot, Noverre, and others recommended and content himself with the stale repetition of the learnt. Lebrun's typology of passionate expression was disliked by some eighteenth-century commentators for just this reason, that it encouraged manneredness and a departure from naturalism. Furthermore, de Jaucourt's *Encyclopédie* article on 'passion, peinture' regretted the loss of true expressiveness in the faces of real-life French citizens, objecting that they had become over-civilized and artificial. On the other hand, the same article counselled artists still to turn to Lebrun for models in the depiction of sentiment via the physiognomy.

Talma remarked of Lekain that the latter's action on stage was free from the manneredness of most of his contemporaries, 'que les Marcel du temps enseignaient à leurs élèves en les initiant aux beautés du menuet' (*Réflexions sur Lekain*, p. viii). But Talma's father reportedly advised his son to take dancing lessons, telling him in 1788 that 'Je ne vois à Paris que M. Vestris le père qui puisse vous donner ces beaux bras et toutes les grâces du corps dont vous avez besoin.'[10] Objections to manneredness in action rarely implied that the

---

[10] See A. Augustin-Thierry, *Le Tragédien de Napoléon, François-Joseph Talma* (Paris, 1942), p. 23.

critic rejected rules and prescriptions out of hand. It was all a question of degree. In his *Essais sur la peinture* Diderot complains about the effect on students of the time spent in the Académie school learning to draw from a model: this, for him, epitomized the stultifying weddedness to precept, rather than encouragement to observe reality directly, that vitiated academic methods of training. It is during the seven painful years of such study, he claims, that pupils acquire 'la manière' in drawing. The consecrated academic positions are expressive only of constraint and artificiality, and bear no relation to the attitudes and actions of real life (p. 669). A crowning absurdity, he writes, would be for pupils to be sent at the end of their seven years to learn graceful posture and movement from the ballet masters Marcel and Dupré (p. 670). It was precisely this lack of naturalness that Diderot stigmatized in the 'grand' acting style of his day. Imitation is a guiding principle for both the live model at the Académie and the actor, he writes, but the latter should derive his bodily eloquence from life rather than the rule-book:

> Qu'ont de commun l'homme qui tire de l'eau dans le puits de votre cour, et celui qui, n'ayant pas le même fardeau à tirer, simule gauchement cette action, avec ses deux bras en haut, sur l'estrade de l'école? Qu'a de commun celui qui fait semblant de se mourir là, avec celui qui expire dans son lit, ou qu'on assomme dans la rue? Qu'a de commun ce lutteur d'école avec celui de mon carrefour? Cet homme qui implore, qui prie, qui dort, qui réfléchit, qui s'évanouit à discrétion, qu'a-t-il de commun avec le paysan étendu de fatigue sur la terre, avec le philosophe qui médite au coin de son feu, avec l'homme étouffé qui s'évanouit dans la foule? Rien, mon ami, rien. (p. 669)

The extent to which a discipline could be brought under the governance of rules had seemed in seventeenth-century France to be an important factor in deciding whether it deserved academic status or not. (The royal academy of painting and sculpture was founded in 1648, followed in 1661 by the royal academy of dance.) It is equally true that as the eighteenth century wore on, objections to the power exerted by such regulations were made with increasing insistence. In the *Essais sur la peinture* Diderot states emphatically that adherence to rules should never become an obsession. The means of pleasing men aesthetically are infinite, and genius must always be allowed to assert itself over the power of regulation. Rules, through making art a routine, may have been more pernicious than useful (p. 753): they have been of service to the ordinary man, but not to the exceptional one (p. 754).

Discussions about technical competence as against a higher kind of artistic creativity were frequent in the eighteenth century. I have already mentioned their place in the theory of acting, where inspired spontaneity was often contrasted with methodical observation of the rule-book, and the relevance of this debate to the question of the actor's status compared with that of other artists is obvious. Significantly, it can be related as well to discussions about the prestige of different kinds of painting that were still being pursued in the age of Diderot (although they had originated much earlier), and to ancient theories concerning the status of art generally. *Ars* in its classical usage stood in contrast to *ingenium*, and rhetorical treatises had explained their relation to one another. *Ars* was the skill that could be learnt by rule or imitation, whereas *ingenium* signified an innate talent that could not (Baxandall, p. 15). Paradoxically, with the development of academies from the Renaissance onwards it was commonly stated that the academy itself was the guardian of *ingenium*, and that the guild was the home of 'maîtrise', mastery of a purely technical kind. The painter Chardin, who became an Academician, had begun his career as a member of a guild, the Académie de Saint-Luc; and although Diderot stated that his genre was as deserving of honour as that of historical painting, he wrote of Chardin that his kind of art called only for study and patience, no verve and little genius, but much technique and truth (*Salons*, II. 111): the craftsman's skill rather than the artist–genius's divine enthusiasm.

Academies, then, were initially seen as the proper milieu for fostering *ingenium* rather than *ars*. But critics who objected to their hegemony saw their rigid structure and their assumption that everything to do with the promotion and appreciation of art could be brought within the scope of rational understanding and logical precept as inimical to the free development of genius. Although the real threat to academies came with the Romantic view of the artist as one who produced masterpieces by the light of inspiration rather than through technical knowledge and academic rule, artists like David were complaining about these institutions in the Revolutionary years[11] just as Diderot and others had done earlier in the century. David and fellow-rebels eventually seceded from the Académie royale de peinture et de

---

[11] See David's letter to the Assemblée nationale on the Académie royale de peinture et de sculpture, in *Deloynes*, LIII, no. 1513; and, on the general question of abolishing academies, *Procès-verbaux du Comité d'instruction publique de la Convention nationale*, II. 242 ff.

sculpture, but in 1793 the institution itself was closed along with all the other royal academies.

Lekain, Talma, and other actors insisted on their right to retain a high degree of personal freedom in their performing style, and thought that their prolonged study not just of acting, but of other academic subjects too, was proof of their entitlement to do so. Yet they also subscribed to the ideals of academic training that were formulated from mid-century on, Lekain as an architect of acting academies and Talma as one of their products. These contrasting attitudes indicate that in the acting profession regulation and independence might coexist. The old debate between the upholders of method and of inspiration in performance, associated with François Riccoboni and Rémond de Sainte-Albine respectively, could be resolved.[12]

When, in 1764, d'Hannetaire noted that there had been talk of establishing acting schools, he justified his view that actors should be granted the same official recognition as painters with an appeal to the notion of unity in the arts: 'l'art du comédien, ainsi que celui des auteurs qui travaillent pour le théâtre, [est] à peu près le même que celui du peintre, *ut pictura poesis*, les uns et les autres ayant toujours la nature pour objet' (p. 36). In the fifteenth century, as we have seen, a belief in correspondence between the arts had been made part of the claim that painting deserved 'liberal' status. In particular, theorists had stated that the painter's prime concern was the arousing of the passions, as it was the orator's, and as writers on stagecraft were later to argue with respect to the actor. Alberti's *Della pittura*, which declared that history painting moved the spectator when the characters depicted exhibited his own emotions, attributed importance to the artistic rendering of emotion through gesture and facial expression. Nature, Alberti wrote, would teach the painter those bodily motions that revealed an inward state of passion. He consequently advised artists to attend to human life around them, as actors and dancers were later counselled to observe the world of men for lessons in the representation they attempted on stage. Leonardo particularly recommended the study of the dumb, whose movements must possess particular expressiveness in virtue of the fact that gestural language was the only type available to them. Towards the end of the sixteenth century Lomazzo discussed in the book on motion of his *Trattato dell'arte della pittura* (1584) the physical movements to which different

---

[12] See the former's *L'Art du théâtre* (1750) and the latter's *Le Comédien* (1747).

emotions of the soul gave rise, and attempted a classification of all emotions together with the gestures and expressions by which they were revealed. The direct effect of visual images on the beholder's sensibility could then be incorporated within the rhetorical theory of painting, as it was later to be into the theory of acting.

Lebrun's Académie lectures on expression, and the accompanying illustrations (published in 1698), provided a pattern-book whose influence continued into the eighteenth century and beyond.[13] But in painting, as in pantomime, its implications were sometimes resisted, or indirectly called into question by disagreement among critics as to what a particular expression betokened. Dubos's *Réflexions* had pertinently observed both that the meaning of particular gestures varies from one country to another and that some gestures have a purely conventional significance.

Diderot's work illustrates, inadvertently as well as by design, the potential for such controversy. An unwitting illustration is provided by his celebrated commentary on a painting of a girl weeping over a dead bird which Greuze exhibited in 1765. Diderot concludes from the girl's facial expression that she is really crying out of remorse: she supposedly caused the bird's death because she neglected it while yielding to a lover.[14] But contemporary critics took a different view (as Diderot notes, *Salons*, II. 147), declaring that the girl was simply expressing a child's grief at intimations of mortality. Mathon de la Cour plausibly considers her to be eleven or twelve, 'âge où le besoin d'aimer fait qu'on se livre au premier objet qui se présente' (ibid., p. 35); Diderot, on the other hand, writes that while her head makes her look fifteen or sixteen, her hand and arm are of an eighteen- or nineteen-year-old (p. 147). Significantly, in the *Pensées détachées sur la peinture* Diderot expresses the wish that a symbol for remorse might be invented for painters and other artists to use (p. 769). If it had been, it would have added to the fund of symbols into which painters had long dipped for the purpose of expressing meaning in a medium whose non-temporal nature often made such expression difficult. Greuze himself made ample use of symbols betokening lost virginity, such as

---

[13] See Jennifer Montagu, 'Charles Le Brun's *Conférence sur l'expression générale et particulière*', Ph. D. (London, 1959); John Montgomery Wilson, *The Painting of the Passions in Theory, Practice and Criticism in Later Eighteenth-Century France* (New York/London, 1981); and B. R. Tilghman, *The Expression of Emotion in the Visual Arts. A Philosophical Enquiry* (The Hague, 1970).

[14] Diderot was probably aware that certain Latin poets had identified the penis with a bird. See J. N. Adams, *The Latin Sexual Vocabulary* (London, 1982), pp. 32–3.

cracked eggs, cracked mirrors, and cracked pitchers, in his own strongly narrative art.

Facial expression, then, could both aid and mislead the beholder in determining 'meaning' in the visual arts. In the *Salons* Diderot berates the artist Hallé for his ineptness at conveying appropriate expression of this kind in an allegorical painting of the city fathers of Paris welcoming Minerva and Peace to their company. He remarks that the provost's air of inanely benevolent welcome suggests he is inviting the two visitors to take a cup of chocolate (*Salons*, III. 71). There are many other instances in representational painting of the obscurity, intended or otherwise, which may attend a painter's portrayal of expression. If eighteenth-century French art offers no parallel to the inscrutability of a Mona Lisa, it contains various examples of figures whose aspect carried different meanings for different observers. The facial expressions of Watteau's characters, for instance, are often equivocal in this manner.

A similar ambiguity was sometimes noted by writers on contemporary French acting. Although performers like Mlle Clairon were widely admired for the range of emotions they were able to convey through the variety of their facial expressions, some commentators remarked that such expressions could be hard to interpret. In his *Ideen zu einer Mimik* Engel reported Descartes's observation that some people make the same face when they are crying as others when they are laughing (I. 141). According to Gachet, the physical distinctions between gestures of joy and pain might be fine ones, and only contextual information (more readily available in drama than on the painter's canvas) would settle the question of their exact significance. Certainly it is often difficult for the modern observer to determine from the depiction alone whether the 'passionate' expression on a character's face in a painting is intended to convey happiness or misery, joy or pain; and this is even more true of the illustrations showing 'têtes des passions' which accompanied Lebrun's Academy lectures. The mechanistic interpretation of action and expression, in other words, was often felt to be unreliable. This uncertainty might also derive from the impossibility, in many cases, of determining whether spontaneity or calculation had prompted physical movement and expression. This difficulty was mentioned by some writers in connection with the human physiognomy, and caused them to distance themselves from contemporary physiognomical theories. Diderot's *Le Neveu de Rameau* describes a man whose face is that of an idiot, but whose character is cunning and acute, and, conversely, a fool who looks intelligent.[15] Here Diderot's suggestion is that such impressions are conveyed by the individual's fixed, unchanging features; but clearly the adoption of a misleading facial expression is a possibility too.

The orators of antiquity had commented on the effectiveness of *actio* as a means of expressing emotion, while reserving the greater part of their instruction for the preparation and delivery of verbal discourse. As we have seen, the ancients passed on to later cultures a delight in gestural action in serious eloquence as well as in popular entertainment, and their interest in pantomime was much discussed by writers in the eighteenth century. It had a marked influence on theories of *actio* in the theatre. But so too did developments of classical doctrine by writers on painting. Lebrun made use of ancient poetic theory in his attempt to show how the mute and static art of painting could yet attain the expressiveness of verbal language. He developed a notion of 'péripéties', derived from the *peripateia* which Aristotle describes in the *Poetics*, to indicate how painting might achieve the temporal richness available to the writer by mingling representations of various passions—such as happiness and misery—on the canvas.[16] In the eighteenth century Shaftesbury remarked on the means which painters could employ to overcome the temporal limitations of visual art (Gombrich, op. cit., p. 40 ff.), as in Rubens's Marie de Médicis sequence; and a similar procedure may be discovered in the *inventio* of Watteau's two *Embarquements pour l'île de Cythère*, where each couple in the procession evokes one of the successive episodes in the pursuit of love (Thuillier, p. 204), or in narrative paintings like Greuze's *L'Accordée de village*.

If the art of painting could be elevated by the successful establishing of similarities between it and ancient poetics and oratory, the prestige of acting might be similarly assured if analogies between it and the art of painting were convincingly demonstrated. So it was assumed by many theorists in eighteenth-century France. But their assumption was challenged by others. I have already referred to Dorfeuille's common-sense observation that the creation of exactly

[15] *Le Neveu de Rameau*, p. 59; see also Graeme Tytler, *Physiognomy in the European Novel: Faces and Fortunes* (Princeton, 1982), pp. 140-3.

[16] See Jacques Thuillier, 'Temps et tableau: la théorie des "péripéties"', in *Stil und Überlieferung in der Kunst des Abendlandes*, 3 vols. (Berlin, 1967), III. 203.

similar effects by actor and painter might be artistically regrettable, because the resources of the two arts were incommensurable. The actor's art was one of movement, and the painter's of immobility. In other words, different kinds of 'truth' should be striven for in the different artistic mediums. Dorfeuille was not alone in pointing to the lack of correspondence between visual art and stage performance: many contemporaries saw that the hybrid nature of the latter, its conjunction of word and action, rendered facile comparison misleading. In any case, the contention that drama might be damaged by the introduction of effects from sculpture or painting was often countered by critics of the time with the assertion that visual art, equally, could not necessarily accommodate 'theatrical' or dramatic elements.

Certain eighteenth-century paintings gave rise to critical reflections on this subject. At first sight, an attitude depicted in Vanloo's portrait of Mlle Clairon as Médée suggests that the painter was guided by dramatic rather than painterly conventions. The penultimate version of the picture showed Jason from the back, flagrantly conflicting with a familiar prescription from the world of acting. Intriguingly, in the final version Vanloo duly altered the posture to give a sideways view. Contemporary reports suggest, however, that this alteration was made not for the sake of conforming to theatrical proprieties, but because observers had found the preceding version dull. One commentator writes that 'la figure de Jason, quoique belle, ayant paru à plusieurs personnes peu intéressante, en ce qu'elle était vue par le dos, l'auteur l'a changée ainsi que toute la composition, qui en est devenue plus grande et plus pittoresque; Jason est dans une vue plus agréable.'[17]

It is left for the critic of the *Année littéraire* to voice a strong theoretical objection to the fondness evinced in many eighteenth-century paintings for borrowing effects from the performance of drama. The conventions of the theatre, he declares, are not equivalent to those of painting, and should not be heedlessly applied to the latter art. Most of the criticisms quoted in chapter 3 are based on the belief that theatricality has caused a breach of decorum in a painting or sculpture; but to observe the proprieties governing serious drama, according to the *Année littéraire*'s critic, may be entirely misplaced in a work of visual art. The picture which stimulates his reflection is a work of Ménageot's commissioned by the city of Paris on the occasion of the

[17] MS *Exposition de peintures, sculptures et gravures du Salon* (*Année littéraire*, 1759, in Deloynes, XLVII, no. 1257).

Dauphin's birth, and shows the baby being presented to the officials of the city by the allegorical figure of France. The attitudes of the characters call forth this comment:

Je m'imagine voir des acteurs sur la scène qui n'oseraient tant soit peu tourner le dos aux spectateurs, crainte de leur manquer de respect. Les convenances théâtrales ne sont pas celles de la peinture. Un préjugé gothique contraint l'acteur de se conformer aux lois prescrites par l'usage; mais l'artiste ne connaît point ces ridicules entraves; le Génie, le Goût, la Nature: voilà ses lois, ses guides, ses modèles; M. Ménageot devait les consulter. Au lieu de placer par exemple presque tous les échevins de profil, afin de les faire regarder la France, qui présente le Dauphin, il aurait pu représenter les figures allégoriques sur un plan plus avancé; alors le groupe des officiers municipaux aurait été susceptible de plus de développement, d'harmonie et d'effet; les attitudes seraient plus variées, et chaque personnage offrirait plus de mouvement, de contraste et d'opposition; au lieu que dans ce tableau, toutes les figures paraissent collées sur la toile. (*Année littéraire*, vol. VI, 1783, letter XIII)

Four years later, however, the critic of the *Année littéraire* presents a modified view. Although it is ridiculous, he states, to subject artists to the 'bienséances théâtrales', such as the rule which prohibits an actor's turning his back to the audience, in Robin's painting of St Louis disembarking in Egypt 'n'est-ce point pécher contre les convenances que de représenter le héros, le principal personnage d'un sujet, entièrement vu par derrière?' (*Année littéraire*, vol. VII, 1787, letter XIX). And later in the Salon review in which he takes issue with Ménageot's painting, the same critic castigates Brenet's *Bayard* for allowing one character to sit while Bayard himself stands—a breach of propriety not according to the theatre's rules, admittedly, but by those of polite society.

The lesson of these examples is that the search for 'constants'[18] in the various arts—all that is implied by the eighteenth century's repetition of Horace's 'ut pictura poesis'—is often a vain one. Correspondences there may be, but, as Diderot and others saw, they are rarely of a straightforward kind. The notion of propriety ('decus') was originally a rhetorical concept, and passed from the theory of rhetoric to that of the other arts; but its meaning is separate for each of them. The idea of decorum was often mentioned in eighteenth-century discussions of the stage performer's *actio*. It was frequently observed that the actor is not required to submit to the kinds of

[18] See Basil Munteano, *Constantes dialectiques en littérature et en histoire* (Paris, 1967).

restraint necessary in the pulpit or the law-court, although many writers, and many actors, believed that control was requisite in the latter's eloquence as well as in the preacher's or advocate's, and that the days of the theatrical 'énergumène' were past. 'Decorum' was one of the most pervasive aesthetic concepts of antiquity, and it implied both that the parts of an artistic whole should be congruent with one another, and that the work of art should bear a 'fitting' relationship to the world it reflected. An ideal of imitation in art that subsumed its reflection of the world under the principle of propriety was posited. Cicero approvingly quoted the actor Roscius's statement that a sense of fitness, of what was becoming, was the main thing in art (*De oratore* i. 132).

During the Renaissance, and especially in neo-classical theory, the requirements of decorum could lead to a rigid separation of artistic genres, with 'academic' pronouncements on what was fitting in each. Thus painting, for example, was divided into different types—historical, landscape, genre, and so on—whose various elements were prescribed by a body of rules. Similarly in acting, the style of performance was, at least in theory, determined by the type of play in question. Tragedy called for stylized declamation, both verbal and gestural, that was elevated above the everyday, and for a grand impressiveness that resembled the 'sublime' register of history painting. Comedy, on the other hand, enjoyed a comparative freedom from rules of performance. But in acting, as in painting, the eighteenth century brought a call for the relaxation of many of the prevailing rules. Diderot's stated desire in the *Entretiens sur 'Le Fils naturel'* to see serious acting brought closer to the performing style of more popular modes, and for a degree of 'bodily communication' to be introduced that had previously been reserved for comedy and pantomime rather than serious drama, was echoed in the efforts of painters to give the traditional 'low' style of genre painting the elevation that belonged to the historical type. Diderot himself sympathized with the new aspirations of artists like Greuze to see a relatively humble mode of painting—the moral scene drawn from everyday life—accorded the magnitude in theory that it seemed inherently to possess. At the same time, it must be admitted, Diderot harshly criticized the ineptness of Greuze's one attempt to attain the exalted heights of history painting in his *Sévère et Caracalla* of 1769, declaring that the artist had abandoned his true style without managing to infuse his canvas with the kind of exaggeration that history painting required (*Salons*, IV. 104).

Sévère et Caracalla, in other words, was not 'becoming' according to the established rules. Later in the century the anti-academic school of David successfully imposed its own 'heroic' interpretation of ordinary characters and their lives on episodes that did not conform with those traditionally thought appropriate to history painting.

Through the contention that the mediums of painting and literature are essentially separate from one another, many commentators tried to account for the power that generically different works of art exercise over their audience or beholder. Quatremère de Quincy's view that the amalgamation of different art-forms produces a whole in which the effect of the constituent arts is less than the effect they separately exert has already been discussed. The arts, according to this interpretation (Quatremère is discussing opera), are incommensurable with one another, and cannot be added to other arts to create a *Gesamtkunstwerk* whose value is simply a compound of individual artistic values. The opinion that the arts should remain autonomous in order to make their most powerful appeal clearly diverges from that expressed by Sainte-Albine in connection with drama, which he regarded as a total art-form, and superior in virtue of its composite character to the art of painting. It is tempting to argue that where correspondences were still perceived, this simply reflected the fact that an aesthetic attitude had been brought to the appreciation of each art, and that beyond this they should be seen as ineluctably different.

At the end of an enquiry into the ways acting was taught in eighteenth-century France, it is appropriate to ask whether the art of *actio* was considered to have progressed over the period or not. Diderot's *Lettre sur les sourds et muets* of 1751 observed that the mute play of the actor was rarely deserving of praise, and he devoted a part of the *Entretiens sur 'Le Fils naturel'* to elaborating this opinion. In 1774, similarly, the prince de Ligne wrote that wild gesticulation was still sadly prominent in theatrical performance. Later, Napoleon reportedly told Talma that his performance as Néron in Racine's *Britannicus* was too gestural (Regnault-Warin, p. 498). Early in the nineteenth century Geoffroy also criticized Talma's acting style in this respect, and observed that the schools which Lekain had wanted to see established had done nothing to improve standards of performance: 'l'école a été formée et n'a remédié à rien: les professeurs de déclamation ont été admis à l'Institut, et la déclamation ne vaut pas ce qu'elle valait du temps où les comédiens étaient exclus de l'Académie Française' (*Cours de littérature dramatique*, VI. 177).

It was widely agreed that *actio* was a discouragingly difficult art to perfect. Noverre's *Lettres sur la danse*, for instance, reports the words of an actor who, although moved to tears by Diderot's *Le Fils naturel* and *Le Père de famille*, believed they could not be well performed:

> cette action *pantomime* serait l'écueil contre lequel la plupart des comédiens échoueraient. La scène muette est épineuse, c'est la pierre de touche de l'acteur. Ces phrases coupées, ces sens suspendus, ces soupirs, ces sons à peine articulés demanderaient une vérité, une âme, une expression et un esprit qu'il n'est pas permis à tout le monde d'avoir. (p. 262)

Great actors, Noverre writes, will be of a mind with Diderot; lesser ones will not:

> c'est qu'il [Diderot's new genre] est pris dans la nature; c'est qu'il faut des hommes pour le rendre et non pas des automates; c'est qu'il exige des perfections qui ne peuvent s'acquérir si l'on n'en porte le germe en soi-même et qu'il n'est pas seulement question de débiter, mais qu'il faut sentir vivement et avoir de l'âme. (Ibid.)

Noverre's sentiment returns his reader to the familiar territory of debates about *ars* and *ingenium*, intellectual detachment and empathy, in acting. His dislike for the rule-book should not, however, be allowed to stand as a final verdict on the eighteenth-century achievement in bodily eloquence.

Conclusion

The principal theme of this book—the embodiment of eloquence—may perhaps be further illustrated by the impact which theories of bodily action made on educational policy in the eighteenth century. The teaching of religious orders, especially that of the Jesuits, embraced an incarnational theology which was given practical expression not just in dramatic performance, but in the oratory of the pulpit, the law-court, and the political assembly. Secular influences reaching back to classical antiquity also helped to preserve a tradition of *actio* in education and the professions. But a third element in the contemporary concern with physical movement was the interest in bodily comportment that had originated in sixteenth-century theories of civility. The preoccupation with developing the whole man, which is evident in influential works like Castiglione's *Il Cortegiano*, included consideration of the part played by bodily eloquence; and, as such, it was initially confined to members of a social élite. But in the course of the eighteenth century there was increasing insistence on the need to educate all men in body as well as in mind. The actor's cause benefited from both these factors, first from the 'civilizing' emphasis on bodily graces, and secondly from the newly democratic appeal acquired by arts of bodily action. With the publication of Rousseau's *Émile*, and then with the various projects for educational reform elaborated by the Revolutionary assemblies from 1790 onwards, such action came to be seen as not merely incidental, but essential, to the development of the individual.

Erasmus's *De civilitate morum puerilium* (1530), of which editions were still being published in the eighteenth century, gave widespread currency in the sixteenth to an ideal of civility that was predominantly an 'externum corporis decorum'.[1] Erasmus's instructions about the proper way to conduct oneself physically, many of which recall the observations of classical rhetoricians on bodily carriage, gesture, and facial expression, in turn influenced writers in France, England, Germany, and Italy. His work is dedicated to the son of a prince, and

---

[1] See Norbert Elias, *Über den Prozess der Zivilisation*, 4th edition, 2 vols. (n.p., 1977), I. 68.

its assumption that courtly life involves an important element of show[2] is one that was subsequently given celebrated expression in literature. To take only the example of France, it figures prominently in neo-classical drama, novels like Mme de Lafayette's *La Princesse de Clèves*, and numerous works by moralists. Among the latter should be counted not just writers such as La Rochefoucauld and La Bruyère, but also the authors of treatises designed to instruct their readers in the social graces. The most famous of these is Nicolas Faret's *L'Honnête Homme, ou l'art de plaire à la cour* (1630). Faret argues the need in polite society for speech to be supported and clarified by the effective use of 'l'éloquence du corps', and remarks that one part of this eloquence, action,

> se doit aussi grandement considérer, étant comme elle est l'âme de tous les discours que nous faisons. En effet nos paroles languissent si elles ne sont pas secourues, et l'on voit plusieurs personnes en la bouche de qui les plus belles choses semblent être mortes, ou du moins sont si froides qu'elles ne touchent point; et d'autres savent animer les moindres de tant de grâce qu'elles délectent tous ceux qui les entendent.[3]

Some of Faret's observations closely resemble recommendations later made by theorists of acting. Such are his remarks on what he calls 'contenance' (or 'une juste situation de tout le corps', which forms 'cette bonne mine que les femmes louent tant aux hommes') and facial expression, particularly that of the eyes: 'c'est par eux que notre âme s'écoule bien souvent hors de nous, et qu'elle se montre toute nue à ceux qui la veillent pour lui dérober son secret' (p. 237).

The integration of instruction about bodily comportment into formal treatises on education is exemplified by Rollin's *Traité des études* (published in 1726, but reflecting seventeenth-century pedagogical beliefs). The eighth book of this treatise observes that physical exercise is important for schoolboys,[4] and in so doing continues a line of argument to which Locke's *Some Thoughts Concerning Education* (1693) had given great prominence. (Locke's first section deals with this matter, and conveys his belief that training the body should precede training the mind.)[5] Locke's concern is with the education of gentlemen, not of the bourgeoisie and still less of the common people;

---

[2] See Georges Vigarello, *Le Corps redressé* (Paris, 1978), p. 52.
[3] N. Faret, *L'Honnête Homme, ou L'Art de plaire à la Cour* (Paris, 1630), pp. 234–5.
[4] See Jacques Thibault, *Les Aventures du corps dans la pédagogie française* (Paris, 1977), p. 182.
[5] See Jacques Ulmann, *De la gymnastique aux sports modernes* (Paris, 1965), p. 181.

and many other treatises from the sixteenth to the eighteenth centuries, until the schemes of the Revolutionary assemblies for the education of all, are similarly élitist in their discussions. What is to be inculcated into the young is the savoir-vivre requisite in polite society. But this bias is absent from those works which discuss the matter of bodily exercise not from the point of view of *politesse*, but from that of hygiene. An example of this type is Vandermonde's *Essai sur la manière de perfectionner l'espèce humaine* (1758), which contrasts the aristocratic institution of academies for noblemen, in which such arts as horse-riding and handling weapons are taught, with gymnasia on the ancient Greek model, where all might satisfy the natural human impulse towards movement.[6]

Many seventeenth- and eighteenth-century writers on bodily eloquence state or imply that training in *actio* is a form of 'dressage', not a relaxation from the strains of intellectual effort. But Verdier, the author of a *Cours d'éducation à l'usage des élèves destinés aux premières professions et aux grands emplois de l'état* (1777), observes that physical and mental efforts always affect one another mutually. He writes in connection with gymnastics that 'Suivant les lois de l'union de l'âme et du corps, rien ne peut exercer l'esprit sans exercer le cerveau et les organes des sens; et réciproquement, rien ne peut exercer le corps sans affecter et occuper l'esprit.'[7] In explicitly Christian works, as in treatises on the *actio* of the preacher, the need to control physical movement in accordance with the principles of decorum is recurrently stressed. The ascetic de La Salle's *Les Règles de la bienséance et de la civilité chrétienne, à l'usage des écoles chrétiennes des garçons* (1736) sets the tone for such manuals, with observations of this kind: 'toutes nos actions extérieures, qui sont les seules qui peuvent être réglées par la bienséance, doivent avoir et porter avec elles un caractère de vertu' (Preface). 'External' civility is defined by de La Salle in terms which nevertheless remain close to those of the courtly treatise: this quality implies a controlled form of behaviour born of modesty, respect, and attentiveness towards one's fellows. De La Salle's instructions on the social graces are firmly set in the tradition of worldly intercourse, and civility itself is defined as 'une conduite sage et réglée, que l'on fait

[6] Charles Vandermonde, *Essai sur la manière de perfectionner l'espèce humaine*, 2 vols. (Paris, 1756), I. 115–18. On this topic see also Dominique Julia, *Les Trois Couleurs du tableau noir: La Révolution* (Paris, 1982), p. 241 ff.

[7] J. Verdier, *Cours d'éducation à l'usage des élèves destinés aux premières professions et aux grands emplois de l'état* (Paris, 1777), p. 234.

paraître dans ses discours et dans ses actions extérieures, par un sentiment de modestie, de respect, d'union et de charité à l'égard du prochain, faisant attention au temps, au lieu et aux personnes avec qui l'on converse' (ibid.).

The applicability to actors of instructions like these, whether intended for sacred or lay orators or for members of polite society in general, may not be readily apparent. But as I have suggested, the connections between the actor's eloquence and that of the preacher or advocate were widely accepted, even if that acceptance often reflected no credit on the actor himself.

Some commentators take the comparison further still. An interestingly qualified view is presented in 1738 by Moncrif, in his *Essais sur la nécessité et sur les moyens de plaire*. After opening the discussion by stating that in any person who speaks to another, external grace depends on an accord between what is said and the action accompanying it,[8] he proceeds to relate this observation to actors and 'gens du monde':

de même que l'art des comédiens, supérieurs dans leur profession, est de s'approprier toutes ces actions heureuses, de ne les marquer qu'au degré, qu'à la nuance qui convient le plus exactement au fond du caractère et à la situation actuelle du personnage qu'ils représentent, c'est dans les gens du monde le plus ou le moins de délicatesse d'esprit et de sentiment qui fait que ces actions sont plus ou moins agréables. (pp. 44–5)

Such 'actions convenues' are seen as varying over the different social orders, and distinguishing the man whose extraction is noble and whose education has been 'honnête' from his less privileged fellow (p. 45). Despite his earlier praise, Moncrif notes that the model provided by the actor cannot be a sufficient one for those desirous of perfection in the matter of external grace: 'cette perfection . . . est l'ouvrage de la justesse et de la délicatesse de l'esprit' (p. 45, footnote), and mere mimicry of the possibly unreliable histrion, by implication, cannot suffice.

But for Dorfeuille, the actor is himself the very 'universal man' whose formation was the concern of Castiglione in *Il Cortegiano*.[9] Dorfeuille prescribes the broadest education for men or women who want a career on the professional stage. They must study the working of language with close attention, grasp the fundamental principles of

---

[8] [F.-A.-P. Moncrif,] *Essais sur la nécessité et sur les moyens de plaire* (Paris, 1738), p. 44.

[9] On Castiglione see Fumaroli, p. 30; also Wilfried Barner, *Barockrhetorik* (Tübingen, 1970).

prosody, be able to distinguish the different types of verse, become acquainted with the history of culture and the human mind, learn correct diction, be practised at performing a 'jeu muet', pay attention to gesture and see how it differs in comedy and tragedy, familiarize themselves with rhetorical *topoi* and understand the various types of eloquence. In his ninth book Dorfeuille recalls that theatrical performance was a hallowed art in ancient Greece, and expresses his desire (in year IX) to see actors enjoy a similar regard in contemporary France. The advantages, he contends, would be immense. Deeper study, more profound reflection, and more extensive knowledge in actors would force the government to change its indifference toward the profession into respect. Actors would be celebrated and esteemed, and enjoy the confidence of the mighty (IX. 12). The art of acting, he writes, has never been treated with the seriousness it merits. Circumstances have always made it the object of prejudice or caprice, but in Dorfeuille's view its connections with the arts and sciences, even with ethics, political theory, and philosophy, give it a position of supreme importance in society (IX. 10–11). For 'le parfait comédien serait l'homme universel, il commanderait partout la victoire: les Grecs en ont senti l'importance et l'utilité: aussi les arts ont-ils offert chez eux à l'univers étonné des orateurs, des législateurs, des héros, des poètes, des hommes en tous les genres couronnés de tous les lauriers de la gloire' (IX. 11). Dorfeuille's stated desire is to persuade the man of the theatre and the orator that without instruction they will never have perfect knowledge of their art. The faculty of speech and all the advantages that attend it are insufficient when unaccompanied by the art of persuasion in all its forms. This art of persuasion, according to Dorfeuille, is the principle of universal harmony. When it is united with the word in the theatre, at the Bar, on the podium or in the pulpit, it becomes a javelin thrown by a skilful warrior; but the word on its own is the feeble instrument of a schoolboy (IX. 13).

But what, if anything, had conferred such status on actors as Dorfeuille assumes to be their right even before they undertake his educational programme? Any attempt to answer must remain in part theoretical, as, indeed, is Dorfeuille's assessment of 'le parfait comédien'. He mentions no names and presents no specific human examples, even remarking that he has written his handbook with an eye to the future rather than the present. Somewhat paradoxically, the desire of some commentators to see acting accepted as a liberal art was expressed in terms of that very exclusivity on which the actors of the Comédie-Française had prided themselves throughout most of the eighteenth century, and to which other actors had objected. If the acting profession were to acquire liberal status, it had to be shown that acting was a discipline to which all might not gain free admittance; and, furthermore, that it was not a craft, but an art taught according to principles comparable with those governing other liberal arts. The concern in the second half of the eighteenth century with setting up institutions for the training of actors reflects these perceptions. Appropriately, Dorfeuille was himself behind an effort to found one such institution, as records of a session at the Convention nationale on 11 ventôse an III (1 March 1795) reveal.[10] On this occasion he presented a list of observations on the establishment at his own expense of an Odéon (then a vacant *salle* in the faubourg Germain) and a school of drama. According to Dorfeuille, the necessity of establishing such a school was widely recognized. An earlier report to the Comité d'instruction publique, of 19 pluviôse an III (7 February 1795), records Dorfeuille's belief that the freedom of theatres has been responsible for a decline in acting standards, and made it necessary to impose a formal structure controlling entrance to the profession. This belief is in line with one expressed some years previously by Magne de Saint-Aubin, who remarked that none of the liberal arts—which he listed as painting, sculpture, architecture, advocacy, and medicine—was open to anyone desirous of practising it.[11]

But discussion did not remain concerned exclusively with the professional proficiency of the actor, for his skill at performing was not taken as the sole criterion by which his worth might be judged. It was with the introduction of the belief that the actor must be a good man that arguments about his status developed, and the analogies drawn with Cato's and Cicero's 'vir bonus dicendi peritus' point to a significant theoretical extension of the actor's role. Like the orator of antiquity, he was to be morally good as well as skilled at declaiming or performing; and other observations concerning his resemblance to the orator combined to suggest his equal right to be considered a practitioner of a liberal art.

Before pursuing the connections between the theory of rhetoric and that of acting, it is worth reporting some contemporary observations

[10] Archives nationales, F^{17}1069, dossier 6.
[11] [Jacques Magne de Saint-Aubin,] *La Réforme des théâtres* (Paris, 1787), pp. 91-2.

about the moral character of actors. In 1759 Villaret published his *Considérations sur l'art du théâtre*, dedicated to Jean-Jacques Rousseau and written to defend drama and actors against attacks such as the one contained in Rousseau's *Lettre à M. d'Alembert sur les spectacles*, which had appeared the previous year. Of particular significance is Villaret's concern with demonstrating that the long-standing suspicion of the actor's morals is unfounded. He notes that 'depuis que nous avons des spectacles réguliers en France, jamais comédien n'a été immolé à la sûreté publique, en expiation de ses forfaits. Feuilletez, compulsez les registres criminels, vous ne verrez point leurs noms inscrits dans ces fastes du crime.'[12] This is not, Villaret says, because the authorities are impotent, or actors above the laws. 'Ils [actors] sont tranquilles, ils ne troublent point l'ordre public.' Moreover, Villaret claims, it is the moral elevation of the plays they daily perform that has effected their own moral development, for as 'organes journaliers des plus sublimes leçons de vertu, il n'est pas possible que leur âme n'en acquière le goût: ils se font aimer.' Their company is sought by the discriminating: 'les personnages sensibles aux agréments de la société recherchent leur commerce et cultivent leur amitié; ils sont ordinairement doux et civils' (pp. 74–5). Villaret ends with a panegyric to the humanity and selflessness of the actor:

Dans les provinces ils consacrent, sans y être contraints, des représentations dans le cours de l'année aux besoins de l'indigence. Tel déclamateur outré contre cette profession prétendue profane ne retrancherait pas la moindre portion des revenus, plus que superflus, qui lui sont assignés, en faveur de l'humanité souffrante, tandis qu'un comédien, sans ostentation, apprend à resserrer les bornes de son nécessaire sans autre motif que de remplir les fonctions d'homme sensible. (pp. 75–6)

A few years later, de La Tour, whose beliefs about the theatre echo Rousseau's, declares without satisfaction that the profession of actor, reviled half a century ago, is now revered.[13] Almanacs, 'tablettes', and dictionaries of the theatre are published in profusion, all filled with information about histrions. In 1787 Magne de Saint-Aubin published his work on the *Réforme des théâtres*, which is partly concerned with the need to create a proper centralized administration for running theatres and training actors, and in which he echoes Villaret's

[12] Claude Villaret, *Considérations sur l'art du théâtre dédiées à M. Jean-Jacques Rousseau, citoyen de Genève* (Geneva, 1759), p. 74.
[13] [Bertrand de La Tour,] *Réflexions morales, politiques, historiques et littéraires sur le théâtre*, 20 vols. in 10 (Avignon, 1763–76), I. 3.

observations about the high esteem in which many actors are held in town and country. *Amateurs* of the theatre, enjoying their company, freely invite them to their houses, although some, Saint-Aubin admits, are unworthy of such favour (p. 35).

The status which many claimed for actors as their right was granted *de facto* with the Revolution, although it was not legally decreed until 1849.[14] Millin de Grandmaison noted with approval in his essay *Sur la liberté du théâtre* (1790) that the profession of actor was no longer regarded *a priori* as reprehensible: faced with claims by actors that they be granted civil status, the Assemblée nationale had acceded, and actors had become electors and 'élevés pour la plupart aux grades militaires dans les bataillons de leurs districts.'[15] A line in Aude's play *Le Journaliste des ombres, ou Momus aux Champs-Élysées*, performed in July 1790, was enthusiastically applauded by the audience for its reference to this matter: 'S'il [Lekain] eût vécu plus tard, il mourait citoyen' (Étienne and Martainville, I. 120). Some commentators suggest that certain actors thenceforth spent more time performing civic duties than plays (ibid., p. 140 f.).

In his discussion of the national fête, where he argues that all the people should be involved as actors in the ceremony, de Moy emphasizes that he is not trying to belittle the art of the professional actor. On the contrary, he acknowledges that the successful performer must be possessed of great talent and even virtue. This last point is important, for it allows de Moy to develop the notion that no actor can convey sentiments which he does not himself experience:

Comment la vertu s'exprimera-t-elle par sa bouche, s'il n'en a pas lui-même? Il peut bien répéter les mots, les phrases qu'il a appris: un souffleur est là pour rappeler ce qui échappe à sa mémoire; mais le ton, le geste, le port, l'attitude, tout ce qui doit partir de son fond: sa sensibilité seule peut le lui inspirer. (*Des fêtes*, pp. 29–30)

In recalling to de Moy's reader the familiar notion of the 'sensible' actor who has to be emotionally involved in his part for his performance of it to be convincing, this observation brings to mind the related opinion that there is a necessary connection between moral uprightness and the successful creation of art. Diderot had earlier posited the same link in his discussion of Boucher, whose deplorable style of

---

[14] See André Villiers, *L'Art du comédien* (Paris, 1959), p. 5.
[15] A. L. Millin de Grandmaison, *Sur la liberté du théâtre* (Paris, 1790), pp. 9–10.

painting he saw as the direct consequence of the artist's depraved moral character:

Je ne sais que dire de cet homme-ci. La dégradation du goût, de la couleur, de la composition, des caractères, de l'expression, du dessin, a suivi pas à pas la dépravation des mœurs. Que voulez-vous que cet artiste jette sur la toile? Ce qu'il a dans l'imagination. Et que peut avoir dans l'imagination un homme qui passe sa vie avec les prostituées du plus bas étage? (*Salons*, II. 75)

(At other times, it is true, Diderot states that great art can be born of immoral deeds, such as the crimes engendered by Christianity, and that artistic geniuses may be dispensed from observing conventions of decent behaviour.)[16] The requirement, in other words, is that the artist should be good not just in the professional sense that he acts or paints competently, but also in a permanent (moral) sense. Indeed, the latter is seen as a prerequisite of the former. This is directly in the tradition of Cato's dictum, repeated as it was by Cicero and other orators of antiquity.[17] In the case of the orator, it was argued by the ancients that eloquence was only legitimate when it was associated with goodness in the absolute sense, embodied in the orator, and that 'actio sequitur esse'. For Cicero, as for Cato before him, the good man was one who willingly fulfilled the duties of the citizen in private and in public life. Extended to cover the case of actors, this notion freed the histrion from the imputations laid against him by Plato, followed by Rousseau and others, that he was a mere copyist; for according to the new view he enjoyed an autonomous existence in the social world of men. The requirement for civic goodness may the more readily be applied to the case of the actor in late eighteenth-century France in that he performed a civic duty at the same time as competently exercising his profession, performing plays that gave physical form to moral requirements regarded as appropriate under the new régime.

But what other indications are there that the ancient model of rhetoric was, during this period, taken as a fitting one for actors to observe? There was an established precedent for invoking similarities with the discipline of rhetoric in an effort to prove the status of another art: theorists of painting had done precisely that two centuries earlier. Then, indirectly, there was the attempt made by actors and writers on acting to show that their art was based on the same principles as

[16] Diderot, *Salons*, III. 148–9 and *Le Neveu de Rameau*, pp. 73–6; also Fried, p. 81.
[17] See Alain Michel, *Rhétorique et philosophie chez Cicéron* (Paris, 1960), p. 15 f.

governed that of the preacher or the advocate, and therefore that it was also a respectable one. In this tradition were the arguments advanced about the need for actors to have a profound understanding of men in general and men in particular, as both Socrates and Plato, among others, had stated to be the case for the orator. Such knowledge is Aristotle's *ethos*, the reliability of the speaker, and Cicero's *mores*. According to ancient rhetorical theory, it enables the speaker to arouse the passions of his audience, and thus to persuade it.[18] This last provides a link between rhetorical principles and theories of acting in eighteenth-century France, which often stressed the need to convince the spectator that the moral view advanced in a play was valid. *Actio*, according to both classical rhetoricians and eighteenth-century writers on drama, increased the possibility that such persuasion might be accomplished.

Dubroca describes the attributes desirable in the actor who would inspire men to virtue, fill them with a hatred of vice, and expose the flaws in society (p. 500) as being dignity in comportment, a good memory, intelligence, and a knowledge of mores and characters—all qualities which rhetorical theory had declared to be essential in the orator. For such attributes to become settled in the actor, he continues, he must have a good education. Aristippe confirms the validity of this observation, remarking that the education should be as wide as an orator's or a painter's (p. 37). But it is Dorfeuille who takes the requirements furthest. The aspirant should, quite simply, be schooled in rhetorical theory, because 'La rhétorique forme le goût que le théâtre exige, et les préceptes le dirigent. Le comédien qui veut marcher à la perfection doit faire un cours de rhétorique' (I. 22). The actor's performance is to be governed by the three *officia* of classical rhetoric (although Dorfeuille does not so describe them): to please, to instruct, and to touch (I. 8). According to Dorfeuille, once the actor has learnt that he is an orator, that all levels and types of eloquence belong to him, and that the art of declamation is worthless when it is not supported by the art of persuading, 'il sentira la nécessité et l'importance pour lui de connaître les parties qui composent l'art de persuader ainsi que les moyens qu'il emploie' (IX. 9). But if the actor disdains to study this subject in detail, 'il rampe, il est au-dessous de la perfection pour lui possible' (IX. 10).

---

[18] See Munteano, p. 162; also Klaus Dockhorn, *Macht und Wirkung der Rhetorik* (Bad Homburg, 1968), p. 49.

Implicit in all this, as in other treatises already discussed, is the view that the system of rhetoric provides a model for living. This holds not simply for the actor, but for anyone who aspires to an understanding of his fellow-men, whatever his profession or station in life.[19] For one commentator, writing in 1659, such a system, whether present in rhetoric, poetry, grammar, music, optics, or other mathematical arts, is equally loved by all men because regularity and proportion, 'qui ne se voi[ent] que par l'œil de l'âme', stem from God; and 'c'est enfin par l'unique rapport à cette loi de tous les arts, et à cette loi de l'artisan souverain et tout-puissant, que l'âme humaine rencontre dans soi-même toute beauté et toute forme, tout modèle et tout exemple, toute mesure et toute harmonie.'[20] This observation might be related to the effort I described in the last chapter to show how the actor's art was founded on certain rule-governed principles, as was that of the painter, the dancer, and the orator. For Gibert, writing at the beginning of the eighteenth century, rhetoric is the ordering principle of all men's works, although his discussion is merely of the ordering of words in discourse (I. 10 f.). The precepts of rhetoric are useful to all, he claims, because every man must learn to persuade others of the decency of his ways and intentions (I. ii). Rhetoric teaches us the need for orderliness and restraint in life, and is thus intimately linked with 'les mœurs'. Its applicability to the arts is as clear as its relation to science, for all reveal the 'natural' distribution of matter and forms (II. 48). The theoretical relation of such beliefs as these to the familiar contemporary notion of correspondence between the arts could not be more clearly implied.

A systematic approach to their art, whether or not the system was that provided by rhetoric, was accepted as appropriate by all who believed in the possibility of training individuals for the stage. Molé, Talma's teacher, reportedly advised his pupil to study the treatise by Engel which attempted to reduce to rule the various gestures through which inner states may be expressed. This work, which Molé declared to be little known, formed a bridge between the theories of Diderot, who in the *Paradoxe* 'donne tout à l'art', and Talma himself, 'qui ne voulez rien refuser à la nature' (Regnault-Warin, p. 91). The views expressed here by Molé reflect his belief that art may improve nature,

[19] See R. Chartier, M.-M. Compère, and D. Julia, *L'Éducation en France du XVI⁰ au XVIII⁰ siècles* (Paris, 1976), p. 197.
[20] Preface by des Herminières to Bary, *La Rhétorique française, où l'on trouve de nouveaux exemples sur les passions et les figures etc.* (Paris, 1659).

and has a duty to correct where it perceives faults or weaknesses (ibid., p. 93). Molé's rejoinder to the objection that what is done according to rule will be cold and stiff is also reported by the same witness:

> la règle qui s'offrait d'abord avec clarté à l'esprit se transformera d'elle-même en idée, et se confondra avec le sentiment qui, au besoin, se présentera avec plus de promptitude et de facilité. L'âme, par l'attention qu'elle doit donner à la règle, ne perdra plus rien de sa force, parce que cette attention ne sera plus nécessaire; l'exécution deviendra aussi facile; elle aura autant de vivacité et de souplesse que celle du simple élève de la nature; mais il y aura plus d'effet et plus d'adresse à surmonter les obstacles. (p. 95)

This is valid, Molé adds, for all the arts.

In practical, as opposed to theoretical, terms we still know little about the training of actors in the eighteenth century, and hence about the extent to which rhetorical principles were inculcated in the pupil. Molé's own lectures on the art of the actor, delivered at the Lycée Républicain in an III (1974–5), would no doubt have furnished valuable information, but seem not to have survived.[21] His *Idées jetées au hasard sur l'établissement de l'école* (i.e. the École royale dramatique which existed from 1786 to 1789) mentions that declamation, French prosody, the French language, and dancing will be taught there (Pierre, *Conservatoire*, p. 62 ff.). But Aristippe, writing in 1826, is critical of the teaching offered at the Conservatoire (the successor to the late eighteenth-century acting schools, and at which Aristippe himself became a professor), noting that the repertoire prescribed for pupils was very limited, and that such discussions of acting as existed in print were not stocked in the library (p. 534). This last remark does suggest that many of the theoretical pronouncements available to present-day students of eighteenth-century acting failed to reach the audience for which they were really intended. Furthermore, he notes, for all the high aspirations of those who sought to dignify the art of acting by establishing an academy, the professional life of its former pupils might fall far below the noble ideals of Lekain and others who wanted to preserve the traditions of the Comédie-Française. Talma, admittedly, was a product of the academy, but in its history the Conservatoire has furnished as many actors for the melodramas of Victor Caignez and Pixérécourt as for the plays of Molière and Racine (pp. 532–3).

---

[21] I have failed to discover any trace of them in the various archives of Paris.

Conclusion 173

Such testimony as this indicates that the evidence of the actor's achieving 'academic' respectability towards the end of the eighteenth century must be treated with caution. But against it may be set the fact that after the dissolution of the royal academies and the creation of the Institut de France in 1795, actors were named among the members of the literature and fine arts section. The Directoire appointed Molé as the representative for his profession in this 'classe', with his brother Dallainville a non-resident associate.[22] Aristippe adds the name of Grandmesnil to the list (p. 13), and observes with reference to the institutional dignity gained by the profession that Picard and Alexandre Duval, then (in 1826) members of the Académie française, had once been actors too (ibid.).

There is no doubt that some actors, besides enjoying such status as this, were revered by the public. The case of Molé is a good example. When he fell ill, to the consternation of all Paris, a humorist circulated a poem commenting on the fact that this illness coincided with one suffered by Nicolet's celebrated monkey:

> L'animal, un peu libertin,
> Tomba malade un beau matin.
> Voilà tout Paris dans la peine . . .
> On croit voir la mort de Turenne.
> Ce n'était pourtant que Molé
> Ou le singe de Nicolet.[23]

Throughout this illness Molé's public requested and received daily bulletins on his condition, and a benefit performance was arranged by his fellow-actors to pay his medical expenses. When Molé died, Dallainville asked every member of the Paris theatre audiences to wear a crêpe band (Boissier *et al.*, p. 82). Yet it seems safe to say that such popular adulation was rather for a man who had successfully amused the public over many years than for an actor whose position in society had, through the developments described in this book, become the equal of that enjoyed by the preacher or the orator of antiquity. However intense were the efforts made by politicians of the Revolutionary period to ensure that the theatre should fulfil many functions previously performed by the church, common sense suggests that a high proportion of Frenchmen under the new régime

[22] See G. Boissier, G. Darboux, G. Perrot, G. Picot, H. Ronjon, *L'Institut de France* (Paris, 1907), pp. 80–2.
[23] See e.g. Michèle Beaulieu, 'Le Théâtre et la sculpture française au XVIIIe siècle', *Le Jardin des arts*, 13 (1956), p. 169.

attended the theatre in the legitimate pursuit of simple pleasure—an ordinary inclination whose importance even the planners of the Revolutionary fête acknowledged. Besides, when the Institut de France was reorganized by the 'arrêté' of 23 January 1803, and four classes were substituted for the original three, the art of declamation was no longer allowed its representatives. No explicit reason for this exclusion was given, but in his introduction to the decree which excluded them Chaptal declared that he found it scandalous to see actors sitting next to physicians, geometers, magistrates, poets, and playwrights in the Collège des Quatre Nations.[24]

Evidence suggesting that in the course of the eighteenth century actors attained a much higher status than they had previously enjoyed is thus balanced by a number of counter-arguments. It is true that their excommunication was lifted *de facto* in 1789, that their important role in disseminating sentiments in accordance with Revolutionary ideals was recognized by politicians and other commentators, that they were granted royal academies, and then admitted to the highest 'academy' in the land in the form of the Institut de France, and that many writers ascribed such status to them. But it is also the case that many others denied the right of actors to be more highly regarded than their forebears, that representatives of a minority group (the 'sociétaires' of the Comédie-Française) were unhappy about being reduced to the common level of French actors, and that imputations of 'actorliness' and 'theatricality' in other fields continued up to and beyond the end of the century to be generally disparaging, whether they alluded to the visual arts, to the conduct of performers and spectators at the political assembly, or to that of preachers or lawyers.

Even after 1789 actors were still suspect to the authorities, and claims about their probity were treated with some scepticism. The acts of the Comité de salut public reveal that in 1794 national agents were to be appointed to keep surveillance over the Opéra and the Théâtre de l'Égalité, and particularly to observe the 'conduite publique, morale et politique' of the actors there. Whether or not such measures as these were justified by events, or simply reflected the legacy of mistrust passed on by earlier periods, it is impossible to say. Whatever the case, one powerful reason suggests itself for the expulsion of the actor from the Institut de France, which would otherwise appear a grave indignity, and it is quite unconnected with the ques-

---

[24] See Jeanne Laurent, *Arts et pouvoirs en France de 1793 à 1981* (Saint-Étienne, 1982), p. 27.

tion of morality or politics. No other member of the literature and fine arts section was a mere performer, even where his art involved performance. The musicians elected to membership were composers first and foremost, and only incidentally players; and the other members of the 'classe' were painters, sculptors, and architects, to none of whose professions the notion of performance applies. Of all the representatives, only the actor was not an *auctor*, an originator. The old belief persisted that drama was primarily an art of the word, and only secondarily one of *actio*, and this meant that the actor could not argue his individuality and creativeness in the use of gesture to be of surpassing artistic importance.

The most lucid of eighteenth-century mimes, Diderot's Neveu de Rameau, acknowledges as much. His life of mimicry has made him someone who cannot create, but only copy, and who will never achieve the artistic greatness of an original spirit. When the philosopher Moi taxes him with his failure in artistic production, Lui's only reply is to refer to his invention of gestures and attitudes:

par exemple, l'attitude admirative du dos dont je vous ai parlé; je la regarde comme mienne, quoiqu'elle puisse peut-être m'être contestée par des envieux. Je crois bien qu'on l'a employée auparavant; mais qui est-ce qui a senti combien elle était commode pour rire en dessous de l'impertinent qu'on admirait! (p. 53)

He later tells Moi that his claim to originality lies in his moral turpitude ('J'ai voulu . . . vous arracher l'aveu que j'étais au moins original dans mon avilissement', p. 76), although Moi is unwilling to concede that moral life may be discussed in terms conventionally reserved for the appreciation of art. The gestures with which the Neveu attempts to recreate the artistic productions of others are illusory, in the sense that his musical performances are mute, merely mimed. But Diderot's examination of the world inhabited by this character reveals that the requirement to act a part—which removes individuality from the performer—extends to everyone, and that all assume attitudes in their relations with their fellow-men (pp. 104–6). Towards the end of their conversation Moi asks the Nephew the meaning of the word 'position', after the latter has described the needy man who spends his life adopting and executing them. The Nephew refers his interlocutor to Noverre (in choreography 'position' indicated one of the different ways of placing the feet relative to each other), but adds that the world shows far more attitudes than the dancer's art can display (p. 104). In taking up such 'positions', it is suggested, people forfeit the claim to autonomy which the non-actor, the genius whose freedom from social constraints the two men have earlier discussed, is able to preserve. The Nephew concludes that the conventions of *actio* governing the lives of most humans, even the most powerful in the land, make them puppets; only the great artist, who is not a mere performer, is free and inimitable.

Select Bibliography

Ah! Ah! Encore une critique du Salon! Voyons ce qu'elle chante (n.p., n.d.).
Almanach général de tous les spectacles de Paris et des provinces, 2 vols. (1791 and 1792).
Anecdotes curieuses et peu connues sur différents personnages qui ont joué un rôle dans la Révolution (Geneva, 1793).
Apologie du goût français relativement à l'opéra (n.p., 1754).
Aristippe [Bernier de Maligny], *Théorie de l'art du comédien, ou Manual théâtral* (Paris, 1826).
[Antoine Arnauld,] *Réflexions sur l'éloquence des prédicateurs* (Amsterdam, 1695).
[Arnould-Mussot,] *Almanach forain, ou Les Différents Spectacles des boulevards et des foires* (Paris, 1773).
Gustave Attinger, *L'Esprit de la Commedia dell'arte dans le théâtre français* (Paris, 1950).
Charles Aubertin, *L'Éloquence politique et parlementaire en France avant 1789* (Paris, 1882).
François Hédelin, abbé d'Aubignac, *La Pratique du théâtre*, 3 vols. (Amsterdam, 1715).
A. Augustin-Thierry, *Le Tragédien de Napoléon, François-Joseph Talma* (Paris, 1942).
F.-A. Aulard, *Le Culte de la raison et le culte de l'être suprême (1793–1794)* (Paris, 1892).
—— *Les Orateurs de l'Assemblée Constituante* (Paris, 1882).
—— (ed.), *Recueil des actes du Comité de salut public*, 28 vols. (Paris, 1889–1951).
Gilbert Austin, *Chironomia, or A Treatise on Rhetorical Delivery*, ed. Mary Margaret Robb and Lester Thonssen (Carbondale and Edwardsville, 1966).
[Louis Petit de Bachaumont,] *Mémoires secrets pour servir à l'histoire de la république des lettres en France*, 36 vols. (London, 1777–89).
Mikhail Bakhtin, *Rabelais and His World*, trans. Hélène Iswolsky (Cambridge, Mass., and London, 1968).
B. Barère de Vieuzac, *Mémoires*, 4 vols. (1842–4).
Wilfried Barner, *Barockrhetorik* (Tübingen, 1970).
Dene Barnett, 'The Performance Practice of Acting: The Eighteenth Century', *Theatre Research International*, new series, II (1976–7); III (1977–8); V (1979–80); VI (1980–1).
—— 'La Rhétorique de l'Opéra', *XVIIe Siècle*, 133 (1981).
—— 'Die Schauspielkunst in der Oper des 18. Jahrhunderts', *Hamburger Jahrbuch für Musikwissenschaft*, 3 (1978).

Baron, *Lettres et Entretiens sur la danse ancienne, moderne, religieuse, civile et théâtrale* (Paris, 1824).
René Bary, *Méthode pour bien prononcer un discours et pour le bien animer* (Paris, 1679).
—— *Nouveau Journal de conversations sur toutes les actions publiques des prédicateurs* (Paris, 1675).
—— *La Rhétorique française, où l'on voit de nouveaux exemples sur les passions et les figures etc.* (Paris, 1659).
Michael Baxandall, *Giotto and the Orators* (Oxford, 1971).
Michèle Beaulieu, 'Le Théâtre et la sculpture française au XVIIIe siècle', *Le Jardin des arts*, 13 (1956).
Annie Becq, 'Expositions, peintres et critiques: vers l'image moderne de l'artiste', *Dix-huitième Siècle*, 14 (1982).
Yves-Marie Bercé, *Fête et révolte: des mentalités populaires du XVIe au XVIIIe siècles* (Paris, 1976).
Louis Bergeran, 'Évolution de la fête révolutionnaire: chronologie et typologie', in *Les Fêtes de la Révolution* (see below).
Gösta M. Bergman, 'La Grande Mode des pantomimes à Paris vers 1740 et les spectacles d'optique de Servandoni', *Recherches théâtrales*, 2 (1960).
Jacob Bernays, 'Aristotle on the Effect of Tragedy', *Articles on Aristotle, 4: Psychology and Aesthetics*, ed. Jonathan Barnes, Malcolm Schofield, and Richard Sorabji (London, 1979).
Marie-Louise Biver, *Fêtes révolutionnaires à Paris* (Paris, 1979).
Yvonne Boerlin-Brodbeck, *Watteau und das Theater* (Basel, 1973).
Marie-Jacques-Amand Boïeldieu, *De l'influence de la chaire, du théâtre et du barreau dans la société civile, et de l'importance de leur rétablissement sur des bases qui puissent relever en France leur ancienne et véritable splendeur. Ouvrage politique et moral* (Paris, an XII/1804).
G. Boissier, G. Darboux, G. Perrot, G. Picot, H. Ronjon, *L'Institut de France* (Paris, 1907).
Didier B*** [Boissieu], *Réflexions sur la festomanie qui nous ont été laissées en partant par Robespierre, Chaumette, Pache, Saint-Just, Hébert et autres philosophes de la même volée* (Paris, n.d.).
François-Antoine, comte de Boissy d'Anglas, *Essai sur les fêtes nationales, adressé à la Convention nationale* (Paris, 12 messidor an II/30 June 1794).
—— *Quelques idées sur les arts, sur la nécessité de les encourager, sur les institutions qui peuvent en assurer le perfectionnement, et sur divers établissements nécessaires à l'enseignement public, adressées à la Convention nationale et au Comité d'instruction publique* (Paris, n.d.).
de Boizi [Charles Palissot], *Considérations importantes sur ce qui se passe depuis quelques jours, au prétendu Théâtre de la Nation, et particulièrement sur les persécutions exercées contre le Sieur Talma* (Paris, 1790).

E. Bonnardet, 'Un Oratorien et un grand peintre', *Gazette des beaux-arts*, I (1938).
Philippe Bordes, 'Jacques-Louis David's *Serment du Jeu de Paume*: Propaganda without a Cause?', *Oxford Art Journal*, 3 (1980).
Jacques-Bénigne Bossuet, *Maximes et réflexions sur la comédie*, 4th edition (Paris, 1930).
[P. Bourdelot and] Bonnet, *Histoire générale de la danse sacrée et profane* (Paris, 1732).
Louis Fauvelet de Bourrienne, *Mémoires sur Napoléon, le Directoire, le Consulat, l'Empire et la Restauration*, 10 vols. (Paris, 1829).
Frank Paul Bowman, 'Le "Sacré-Cœur" de Marat (1793)', in *Les Fêtes de la Révolution* (see below).
Brazier, *Histoire des petits théâtres de Paris*, new edition, 2 vols. (Paris, 1838).
Anita Brookner, *Jacques-Louis David* (London, 1980).
Else Marie Bukdahl, *Diderot, critique d'art*, 2 vols. (Copenhagen, 1980 and 1982).
Edmund Burke, *Reflections on the Revolution in France*, ed. F. G. Selby (London, 1890).
Cahier. Plaintes et doléances de MM. les Comédiens Français (n.p., 1789).
Louis de Cahusac, *La Danse ancienne et moderne, ou Traité historique de la danse*, 3 vols. (The Hague, 1754).
Jean-François Cailhava de l'Estendoux, *Les Causes de la décadence du théâtre* (Paris, 1807).
—— *De l'art de la comédie*, 4 vols. (Paris, 1772).
Émile Campardon, *Les Spectacles de la foire*, 2 vols. (Paris, 1877).
Carl Friedrichs von Baden brieflicher Verkehr mit Mirabeau und Du Pont, ed. Carl Knies, 2 vols. (Heidelberg, 1892).
Marvin Carlson, *The Theatre of the French Revolution* (New York, 1966).
[Louis Charpentier,] *Causes de la décadence du goût sur le théâtre* (Paris, 1768).
R. Chartier, M.-M. Compère, and D. Julia, *L'Éducation en France du XVIe au XVIIIe siècles* (Paris, 1976).
François-René, vicomte de Chateaubriand, *Mémoires d'outre-tombe*, 6 vols. (Paris, 1849–50).
G. Chaussinand-Nogaret, *Mirabeau* (Paris, 1982).
[Marie-Joseph Chénier,] *De la liberté du théâtre en France* (n.p., n.d.).
Jacques Chesnais, *Histoire générale des marionnettes* (Paris, 1947).
Sylvie Chevalley, 'Les Bals de la saison d'hiver en 1716–1717', *Comédie-Française*, 66 (1978).
Marie-Françoise Christout, *Le Merveilleux et le 'théâtre du silence'* (The Hague and Paris, 1965).
Hippolyte Clairon, *Mémoires*, new edition (Paris, 1822).
Charles Collé, *Journal et mémoires*, ed. Honoré Bonhomme, 3 vols. (Paris, 1868).

[Charles Compan,] *Dictionnaire de danse* (Paris, 1787).
MS *Compte rendu de l'affaire des auteurs* (Archives nationales, O¹ 843 B).
Alfred Copin, *Talma et la Révolution* (Paris, 1887).
Pierre Corneille, *Writings on the Theatre*, ed. H. T. Barnwell (Oxford, 1965).
Correspondance dramatique entre MM. Mercier (de l'Institut), Cubières-Palmézeaux, auteur dramatique, et M. Simon, avocat, et secrétaire du Comité de lecture de l'Odéon (Paris, 1810).
Jacques-Michel Coupé, *Des fêtes en politique et en morale* (Paris, n.d.).
Louis Courajod, *L'École royale des élèves protégés* (Paris, 1874).
Charles-Maurice Couyba, *Le Parlement français* (Paris, 1914).
Antoine Coypel, *Discours prononcés dans les conférences de l'Académie royale de peinture et de sculpture* (Paris, 1721).
Thomas Crow, 'The Oath of the Horatii: Painting and Pre-Revolutionary Radicalism in France', *Art History*, 1 (1978).
Jules David, *Le Peintre Louis David* (Paris, 1880).
De l'organisation des spectacles de Paris (Paris, 1790).
Deloynes, Collection, 63 vols. (Bibliothèque nationale, Paris).
Deux Siècles d'opéra français (catalogue to exhibition at Bibliothèque nationale, Paris, 1972).
Denis Diderot, *De la poésie dramatique*, in *Œuvres esthétiques*, ed. Paul Vernière (Paris, 1968).
—— *Éléments de physiologie*, ed. Jean Meyer (Paris, 1964).
—— *Entretiens sur 'Le Fils naturel'*, in *Œuvres esthétiques*.
—— *Essais sur la peinture*, in *Œuvres esthétiques*.
—— *Lettre sur les sourds et muets*, ed. Paul Hugo Meyer, in *Diderot Studies*, VII (1965).
—— *Le Neveu de Rameau*, ed. Jean Fabre (Geneva, 1963).
—— *Œuvres complètes*, ed. Jules Assézat and Maurice Tourneux, 20 vols. (Paris, 1875–7).
—— *Paradoxe sur le comédien*, in *Œuvres esthétiques*.
—— *Pensées détachées sur la peinture*, in *Œuvres esthétiques*.
—— *Salons*, ed. Jean Seznec and Jean Adhémar, 4 vols. (Oxford, 1957–67).
—— and Falconet, *Le Pour et le contre*, ed. Yves Benot (Paris, 1958).
Joseph-Antoine-Toussaint Dinouart, *L'Éloquence du corps, ou l'action du prédicateur*, 2nd edition (Paris, 1761).
Discours et motions sur les spectacles (Paris, 1829).
Klaus Dockhorn, *Macht und Wirkung der Rhetorik* (Bad Homburg, 1968).
Claude-Joseph Dorat, *La Déclamation théâtrale, poème didactique en trois chants, précédé d'un discours* (Paris, 1766).
Prosper Dorbec, 'Les Premiers Contacts avec l'atelier du peintre dans la littérature moderne', *RHLF*, XXVIII (1921).
[Dorfeuille, i.e. P.-P. Gobet,] *Les Éléments de l'art du comédien, considéré dans chacune des parties qui le composent, à l'usage des élèves et des amateurs du théâtre*, 9 vols. (Paris, an VII–an IX).

Select Bibliography

David L. Dowd, 'Art and the Theatre during the French Revolution: The Rôle of Louis David', *The Art Quarterly*, 23 (1960).

—— 'Art as National Propaganda in the French Revolution', *The Public Opinion Quarterly*, 15, no. 3 (1951).

—— *Pageant-Master of the Republic: Jacques-Louis David and the French Revolution* (Nebraska, 1948).

Victor Du Bled, *La Comédie de société au dix-huitième siècle* (Paris, 1893).

Jean-Baptiste Dubos, *Réflexions critiques sur la poésie et sur la peinture*, 2 vols. (Paris, 1719).

Jean-François Dubroca, *Discours sur divers sujets de morale et sur les fêtes nationales* (Paris, an VII).

Louis Dubroca, *Principes raisonnés sur l'art de lire à haute voix, suivis de leur application particulière à la lecture des ouvrages d'éloquence et de poésie* (Paris, 1802).

Jean Dubu, 'Bossuet et le théâtre: un silence de l'Évêque de Meaux', *Journées Bossuet: La Prédication au XVIIe siècle*, ed. Thérèse Goyet and Jean-Pierre Collinet (Paris, 1980).

[Ducoudray,] *Correspondance dramatique*, 2 vols. (Paris, 1778).

—— *Il est temps de parler, et Il est temps de se taire, précédés de la lettre au public sur l'établissement d'une école dramatique, protégée par les Comédiens Français*, 2 vols. in 1 (Paris, 1779).

Dumont, *Le Désœuvré mis en œuvre*, in [Nougaret,] *La Littérature renversée, ou L'Art de faire des pièces de théâtre sans paroles* (Paris, 1775).

—— *Le Vol le plus haut, ou L'Espion des principaux théâtres de la capitale* ('Memphis', 1784).

Étienne Dumont, *Souvenirs sur Mirabeau et sur les deux premières Assemblées législatives*, ed. J. Bénétruy (Paris, 1951).

Guillaume Du Vair, *De l'éloquence française (1594)*, ed. René Radouant (Paris, n.d.).

Alexandre Duval, *Réflexions sur l'art de la comédie* (Paris, 1820).

Émile Duval, *Talma. Précis historique sur sa vie, ses derniers moments et sa mort* (Paris, 1826).

Jean Ehrard, 'Les Lumières et la fête', in *Les Fêtes de la Révolution* (see below).

Norbert Elias, *Über den Prozess der Zivilisation*, 4th edition, 2 vols. (n.p., 1977).

J.-J. Engel, *Ideen zu einer Mimik*, 2 vols. (Berlin, 1785-6).

Entretiens sur l'état actuel de l'Opéra de Paris (Amsterdam, 1779).

Essai sur l'opéra (Paris, 1772).

Charles-Guillaume Étienne *et al.*, *Almanach général de tous les spectacles de Paris et des provinces*, 2 vols. (Paris, 1791-2).

Charles-Guillaume Étienne and Alphonse-Louis-Dieudonné Martainville, *Histoire du Théâtre Français depuis le commencement de la révolution jusqu'à la réunion générale*, 3 vols. (Paris, an X/1802).

Exposé de la conduite et des torts du Sieur Talma envers les Comédiens Français (Paris, 1790).

MS *Exposition de peintures, sculptures et gravures du Salon* (*Année littéraire*, 1759, in *Collection Deloynes*, XLVII, no. 1257).

Nicolas Faret, *L'Honnête Homme, ou L'Art de plaire à la cour* (Paris, 1630).

Charles-Simon Favart, *Mémoires et correspondance littéraires, dramatiques et anecdotiques*, 3 vols. (Paris, 1808).

[Charles-Élie de Ferrières,] *Mémoires pour servir à l'histoire de l'Assemblée Constituante et de la révolution de 1789*, 3 vols. (Paris, an VII).

Les Fêtes de la Révolution: Colloque de Clermont-Ferrand, juin 1974 (Paris, 1977).

Feu Séraphin: Histoire de ce spectacle 1776–1870 (Lyon, 1875).

Joseph-Abraham Bénard Fleury, *Mémoires de Fleury de la Comédie-Française*, 6 vols. (Paris, 1835–8).

W. W. Fortenbaugh, *Aristotle on Emotion* (London, 1975).

[Nicolas-Étienne Framéry,] *De l'organisation des spectacles de Paris, ou Essai sur leur forme actuelle* (Paris, 1790).

Peter France and Margaret McGowan, 'Autour du *Traité du récitatif* de Grimarest', *XVIIᵉ Siècle*, 132 (1981).

Michael Fried, *Absorption and Theatricality: Painting and Beholder in the Age of Diderot* (Berkeley, Los Angeles, and London, 1980).

Marc Fumaroli, *L'Âge de l'éloquence* (Geneva, 1980).

Frantz Funck-Brentano, *La Bastille des comédiens. Le For l'Évêque* (Paris, 1903).

Louis Gachet, *Observations sur les spectacles en général, et en particulier sur le Colisée* (Paris, 1772).

Léonard Gallois, *Histoire des journaux et des journalistes de la Révolution*, 2 vols. (Paris, 1845–6).

Roger Garaudy, *Les Orateurs de la Révolution française* (Toulouse, 1939).

Jean-Baptiste-Modeste Gencé, *Vues sur les fêtes publiques et application de ces vues à la fête de Marat* (Paris, an II).

Julien-Louis Geoffroy, *Cours de littérature dramatique*, 6 vols. (Paris, 1825).

—— *Manuel dramatique à l'usage des auteurs et des acteurs* (Paris, 1822).

Balthasar Gibert, *Réflexions sur la rhétorique, où l'on répond aux objections du Père Lamy, bénédictin*, 3 vols. (Paris, 1705–7).

Stanislas Girardin, *Mémoires et souvenirs*, 2nd edition, 2 vols. (Paris, 1829).

Jacques Godechot, *La Grande Nation: L'Expansion révolutionnaire de la France dans le monde, 1789–1799*, 2 vols. (Paris, 1956).

—— *Les Institutions de la France sous la Révolution et l'Empire*, 2nd edition (Paris, 1968).

E. H. Gombrich, '*Icones symbolicae*: The Visual Image in Neo-Platonist Thought', *Journal of the Warburg and Courtauld Institutes*, 11 (1948).

—— 'Moment and Movement in Art', in *The Image and the Eye* (Oxford, 1982).

John Grand-Carteret ed., *L'Histoire, la vie, les mœurs et la curiosité*, 6 vols. (Paris, 1927–8).

Grimm, Diderot, Raynal, Meister, etc., *Correspondance littéraire, philosophique et critique*, ed. Maurice Tourneux, 16 vols. (Paris, 1877–82).

Joseph-François-Louis Grobert, *Des fêtes publiques chez les modernes* (Paris, an X).
Alain-Charles Gruber, *Les Grandes Fêtes et leurs décors à l'époque de Louis XVI* (Paris and Geneva, 1972).
—— 'Les "Vauxhalls" parisiens au XVIIIe siècle', *Bulletin de la Société de l'histoire de l'art français, année 1971* (1972).
Gabriel Guéret, *Entretiens sur l'éloquence de la chaire et du barreau* (Paris, 1666).
Guillaume, *Almanach dansant* (Paris, 1770).
J. Guillaume (ed.), *Procès-verbaux du Comité d'instruction publique de l'Assemblée législative* (Paris, 1889).
—— *Procès-verbaux du Comité d'instruction publique de la Convention nationale*, 7 vols. (Paris, 1891–1907).
Guillemin (maître de danse), *Chorégraphie, ou L'Art de décrire la danse* (Paris, 1784).
Gerhard Anton von Halem, *Paris en 1790*, trans. A. Chuquet (Paris, 1896).
Victor Hallays-Dubot, *Histoire de la censure théâtrale en France* (Paris, 1862).
Jean-Nicolas Servandoni d'Hannetaire, *Observations sur l'art du comédien et sur d'autres objets concernant cette profession en général, avec quelques extraits de différents auteurs et des remarques analogues au même sujet*, 2nd edition (Paris, 1774).
Marie-Jean Hérault de Séchelles, 'Réflexions sur la déclamation', *Magazin encyclopédique, ou Journal des lettres, des sciences et des arts*, I (Paris, 1795).
Robert L. Herbert, *David, Voltaire, 'Brutus' and the French Revolution: An Essay in Art and Politics* (London, 1972).
Roger Herzel, 'Le Jeu "naturel" de Molière et de sa troupe', *XVIIe Siècle*, 132 (1981).
Christel Heybrock, *Jean-Nicolas Servandoni: eine Untersuchung seiner Pariser Bühnenwerke* (Cologne, 1970).
Eugen Hirschberg, 'Die Encyklopädisten und die französische Oper im 18. Jahrhundert', D. Phil. (Leipzig, 1903).
Marian Hobson, *The Object of Art: The Theory of Illusion in Eighteenth-Century France* (Cambridge, 1982).
William Hogarth, *The Analysis of Beauty* (London, 1753).
Kirsten Gram Holmström, *Monodrama, Attitudes, Tableaux Vivants* (Stockholm, 1967).
Seymour Howard, *Sacrifice of the Hero: The Roman Years. A Classical Frieze by Jacques-Louis David* (Sacramento, 1975).
W. G. Howard, 'Ut pictura poesis', *PMLA*, 24 (1909).
W. D. Howarth, *Sublime and Grotesque* (London, 1975).
[Huerne de la Mothe,] *Libertés de la France contre le pouvoir arbitraire de l'excommunication* (Amsterdam, 1761).
Idées sur l'opéra (Paris, 1764).
F. Ingersoll-Smouse, 'Charles-Antoine Coypel', *Revue de l'art ancien et moderne*, XXXVIII (1920).
Robert M. Isherwood, 'Entertainment in the Parisian Fairs in the Eighteenth Century', *Journal of Modern History*, 53 (1981).

Georg Jacob, *Geschichte des Schattentheaters*, 2nd edition (Hanover, 1925).
Aniela Jaffé, 'Symbolism in the Visual Arts', in Carl Jung *et al.*, *Man and his Symbols* (London, 1964).
I. Jamieson, C.-A. *Coypel, premier peintre de Louis XV. Sa Vie et son œuvre artistique et littéraire* (Paris, 1930).
Paul Jarry, 'Notes sur le Colisée', *Bulletin de la Société historique des VIIe et XVIIe arrondissements* (1913).
B. L. Joseph, *Elizabethan Acting* (Oxford, 1950).
Dominique Julia, *Les Trois Couleurs du tableau noir: la Révolution* (Paris, 1982).
Diane Kelder, *Aspects of 'Official' Painting and Philosophic Art 1789–1799* (New York and London, 1976).
Manfred Krüger, *J.-G. Noverre und das „Ballet d'action": Jean-Georges Noverre und sein Einfluss auf die Ballettgestaltung* (Emsdetten, 1963).
Jacques Lablée, *Du théâtre de la Porte Saint-Martin, de pièces d'un nouveau genre, et de la pantomime* (Paris, 1812).
Charles Lacretelle, *Dix Années d'épreuves pendant la Révolution* (Paris, 1842).
Sigismond Lacroix (ed.), *Actes de la Commune de Paris pendant la Révolution*, 15 vols. (Paris, 1894–1914).
Catherine Lafarge, 'L'Anti-fête chez Mercier', in *Les Fêtes de la Révolution* (see above).
Henri Lagrave, *Le Théâtre et le public à Paris de 1715 á 1750* (Paris, 1972).
La Harpe, *Discours sur la liberté du théâtre, prononcé le 17 décembre 1790 à la société des amis de la Constitution* (Paris, 1790).
Gérard de Lairesse, *Le Grand Livre des peintres*, 2 vols. (Paris, 1787).
Bernard Lamy, *De l'art de parler*, 2nd edition (Paris, 1676).
François-Xavier Lanthenas, *Bases fondamentales de l'instruction publique et de toute la constitution libre* (Paris, 1793).
—— *Développement du plan et des motifs de loi ou cadre pour l'institution des fêtes décadaires* (Paris, 16 nivôse an III).
[Joseph de La Porte,] *Les Spectacles de Paris, ou Suite du calendrier historique et chronologique des théâtres* (Paris, 1754–[1778]).
Louis-Marie de La Revellière-Lépeaux, *Essai sur les moyens de faire participer l'universalité des spectateurs à tout ce qui se pratique dans les fêtes nationales* (Paris, an VI).
—— *Réflexions sur le culte, sur les cérémonies civiles et sur les fêtes nationales*, in Dubroca, *Discours* (see above).
Jean-Marie Mauduit-Larive, *Cours de déclamation* (Paris, an XII [1804]).
Pierre Larthomas, *Le Langage dramatique, sa nature, ses procédés* (Paris, 1972).
Jean-Baptiste de La Salle, *Les Règles de la bienséance et de la civilité chrétienne, à l'usage des écoles chrétienne des garçons* (Reims, 1736).
Théodore Lassabathie, *Histoire du Conservatoire impérial de musique et de déclamation* (Paris, 1860).
[Bertrand de La Tour,] *Réflexions morales, politiques, historiques et littéraires sur le théâtre*, 20 vols. in 10 (Avignon, 1763–6).

De L'Aulnaye, *De la saltation théâtrale* (Paris, 1790).
Jeanne Laurent, *Arts et pouvoirs en France de 1793 à 1981* (Saint-Étienne, 1982).
Rensselaer W. Lee, 'Ut pictura poesis: The Humanistic Theory of Painting', *The Art Bulletin*, XXII (1940).
Michel Le Faucheur, *Traité de l'action de l'orateur ou de la prononciation et du geste*, ed. and rev. by Conrart (Paris, 1657).
Le Gras, *La Rhétorique française ou les préceptes de l'ancienne et vraie éloquence accommodés à l'usage des conversations et de la société civile: Du barreau et de la chaire* (Paris, 1671).
Henri-Louis Lekain, *Mémoires* (Paris, 1825).
Népomucène Lemercier, *Du second théâtre français, ou Instruction relative à la déclamation dramatique* (Paris, 1818).
Joseph-Marie Lequinio, représentant, *Des fêtes nationales* (Paris, n.d.).
Antoine de Léris, *Dictionnaire des théâtres de Paris*, 7 vols. (Paris, 1767).
MS *Lettre aux auteurs du 'Journal de Paris'* (*Collection Deloynes*, XVII, no. 435).
Lettre critique à un ami sur les ouvrages de MM. de l'Académie exposés au Salon du Louvre 1759 (Paris, 1759).
Lettre critique de M. le Marquis de *** à M. de Servandoni au sujet du spectacle qu'il donne au Palais des Tuileries (Paris, 1754).
Lettre critique sur notre danse théâtrale (Paris, 1771).
Lettre d'un artiste sur le tableau de Mlle Clairon (Paris, 1759).
Lettre écrite à un ami sur les danseurs de corde et sur les pantomimes qui ont paru autrefois chez les Grecs et chez les Romains et à Paris en 1738 (Paris, 1739).
Lettres à un artiste sur les fêtes publiques (Paris, an IX).
Lettres analytiques, critiques et philosophiques sur les tableaux du Salon (Paris, 1791).
Lettres pittoresques à l'occasion des tableaux exposés au Salon en 1777 (Paris, 1777).
[Jean-Charles Levacher de Charnois,] *Conseils à une jeune actrice* (n.p., 1788).
[Pierre-Alexandre Levesque de la Ravaillère,] *Essai de comparaison entre la déclamation et la poésie dramatique* (Paris, 1729).
George Levitine, 'The Influence of Lavater and Girodet's *Expression des sentiments de l'âme*', *The Art Bulletin*, XXXVI (1959).
[prince de Ligne,] *Lettres à Eugénie sur les spectacles* (Brussels, 1774).
Simon-Nicolas-Henri Linguet, *Annales politiques, civiles et littéraires du 18e siècle*, 19 vols. (London and Paris, 1777–92).
—— *Œuvres complètes*, 2 vols. (Paris, 1779).
Jean Locquin, *La Peinture d'histoire en France de 1747 à 1785* (Paris, 1912).
Robert Lowe, 'Les Représentations en musique au collège Louis-le-Grand de Paris (1689–1762)', *Revue d'histoire du théâtre*, 11 (1959).
[Jacques Magne de Saint-Aubin,] *La Réforme des théâtres* (Paris, 1787).
Charles Magnin, *Histoire des marionnettes en Europe*, 2nd edition (Paris, 1862).
Jacques Mallet du Pan, *Mémoires et correspondance*, ed. A. Sayous, 2 vols. (Paris, 1851).
Malpied (maître de danse), *Traité sur l'art de la danse*, 2nd edition (Paris, n.d.).

P.-J. Mariette, *Abecedario*, ed. P. de Chennevières and A. de Montaiglon, 6 vols. (Paris, 1851–60).
Jean-François Marmontel, *Éléments de littérature*, 3 vols. (Paris, 1879).
—— *Mémoires*, ed. John Renwick, 2 vols. (Clermont-Ferrand, 1972).
Albert Mathiez, *Les Origines des cultes révolutionnaires (1789–1792)* (Paris, 1904).
Mathon de la Cour, *Seconde Lettre à Monsieur ** sur les peintures, les sculptures et les gravures exposées au Salon du Louvre en 1765* (n.p., 1765).
Charles S. Maury, *Essai sur l'éloquence de la chaire*, new edition, 3 vols. (Paris, 1827).
Gita May, 'Diderot and Burke: A Study in Aesthetic Affinity', *PMLA*, 75 (1960).
E. Allen McCormick, '*Poema pictura loquens*: Literary Pictorialism and the Psychology of Landscape', *Comparative Literature Studies*, 13 (1976).
Mémoire sur les danses chinoises, d'après une tradition manuscrite de quelques ouvrages de Confucius (n.p., n.d.).
[Claude-François Ménestrier,] *Des ballets anciens et modernes* (Paris, 1682).
—— *Traité de tournois, joustes, carrousels et autres spectacles publics* (Lyon, 1669).
Louis-Sébastien Mercier, *Du théâtre ou Nouvel Essai sur l'art dramatique* (Amsterdam, 1773).
—— *Paris pendant la Révolution, ou Le Nouveau Paris*, 2 vols. (Paris, 1862).
Joseph Mérilhou, *Essai historique sur la vie et les ouvrages de Mirabeau* (Paris, 1827).
Alain Michel, *Rhétorique et philosophie chez Cicéron* (Paris, 1960).
Aubin-Louis Millin de Grandmaison, *Sur la liberté du théâtre* (Paris, 1790).
Mirabeau à l'Assemblée Constituante (Paris, 1848).
[Honoré-Gabriel de Riqueti, comte de Mirabeau,] *La Lanterne magique nationale* (n.p., n.d.).
—— *La Nouvelle Lanterne magique* (Paris, 1790).
—— *Travail sur l'éducation publique trouvé dans les papiers de Mirabeau l'aîné*, ed. P. J. G. Cabanis (Paris, 1791).
Jean-Baptiste Poquelin de Molière, *Œuvres*, ed. Eugène Despois, 13 vols. (Paris, 1873–1912).
[Francois-Augustin Paradis de Moncrif,] *Essais sur la nécessité et sur les moyens de plaire* (Paris, 1738).
Jennifer Montague, 'Charles Le Brun's *Conférence sur l'expression générale et particulière*', Ph. D., 2 vols. (London, 1959).
—— 'The Painted Enigma and French Seventeenth-Century Art', *Journal of the Warburg and Courtauld Institutes*, 31 (1968).
A. de Montaiglon (ed.), *Procès-verbaux de l'Académie royale de peinture et de sculpture*, 10 vols. (Paris, 1875–92).
John Montgomery Wilson, *The Painting of the Passions in Theory, Practice and Criticism in Later Eighteenth-Century France* (New York and London, 1981).

J. Morange and J.-P. Chassaing, *Le Mouvement de réforme de l'enseignement* (Paris, 1974).

[Charles-Alexandre de Moy,] *Des fêtes, ou quelques idées d'un citoyen français relativement aux fêtes publiques et à un culte national* (Paris, an VII).

Robert Muchembled, *Culture populaire et culture des élites dans la France moderne (XVe-XVIIIe siècles)* (Paris, 1978).

Basil Munteano, *Constantes dialectiques en littérature et en histoire* (Paris, 1967).

Colette Nativel, 'Franciscus Junius et le *De pictura veterum*', *XVIIe Siècle*, 138 (1983).

Charles Nodier, 'Recherches sur l'éloquence révolutionnaire', *Œuvres complètes*, 12 vols. (Paris, 1832–7), VII.

[Pierre-Jean-Baptiste Nougaret,] *La Littérature renversée, ou L'Art de faire des pièces de théâtre sans paroles* (Paris, 1775).

Jean-Georges Noverre, *Lettres sur la danse et sur les arts imitateurs* (Paris, 1950).

—— *Lettres sur les arts imitateurs en général, et sur la danse en particulier*, 2 vols. (Paris, 1807).

Observations critiques sur les tableaux du Salon de l'année 1787 (Paris, 1787).

William Olander, 'French Painting and Politics in 1794: The Great *Concours de l'an II*', *Proceedings of the 10th Convention on Revolutinary Europe, 1750–1850* (1980).

Jean-Jacques Olivier, *Voltaire et les comédiens interprètes de son théâtre* (Paris, 1899).

Mona Ozouf, *La Fête révolutionnaire, 1789–1799* (Paris, 1976).

—— 'Le Simulacre et la fête révolutionnaire', in *Les Fêtes de la Révolution* (see above).

Charles Palissot, *La Critique de 'Charles IX'* (Paris, 1790).

—— *Mémoires pour servir à l'histoire de notre littérature depuis Francois Ier jusqu'à nos jours*, new edition, 2 vols. (Paris, 1803).

La Pantomimanie, contained in the *Répertoire du Théâtre sans Prétention*, 1798–1805 (Bibliothèque historique de la Ville de Paris, 611133).

Jean-Pierre Papon, *L'Art du poète et de l'orateur*, 6th edition (Paris, 1806).

[François and Claude Parfaict,] *Mémoires pour servir à l'histoire des spectacles de la foire, par un auteur forain*, 2 vols. (Paris, 1743).

abbé Parisis, *Questions importantes sur la comédie de nos jours*, 2nd edition (Valenciennes, 1789).

Les Petits Spectacles de Paris (Paris, 1786).

Pierre Peyronnet, 'Le Théâtre d'éducation des Jésuites', *Dix-huitième Siècle*, 8 (1976).

Constant Pierre, *Les Anciennes Écoles de déclamation dramatique* (Paris, 1896).

—— *Le Conservatoire national de musique et de déclamation* (Paris, 1900).

[Roger de Piles,] *L'Idée du peintre parfait* (London, 1707).

Abel Poitrineau, 'La Fête traditionnelle', in *Les Fêtes de la Révolution* (see above).

MS *Portrait de Mlle Clairon, par Carle Vanloo* (*Collection Deloynes*, XLVII, no. 1267).
Charles de Pougens, *Mémoires* (Paris, 1834).
Arthur Pougin, *Acteurs et actrices d'autrefois* (Paris, n.d.).
Bernard Poyet, *Projet de cirque national et de fêtes annuelles* (n.p., n.d.).
Préville, *Mémoires*, in *Mémoires de Préville et de Dazincourt*, ed. M. Ourry (Paris, 1823).
Antoine-François Prévost, *Histoire du chevalier Des Grieux et de Manon Lescaut*, ed. Frédéric Deloffre and Raymond Picard (Paris, 1965).
J.-G. Prod'homme and E. de Crauzat, *Les Menus Plaisirs du Roi, l'École royale et le Conservatoire de musique* (Paris, 1929).
Antoine-Chrysostôme Quatremère de Quincy, *Dissertation sur les opéras bouffons italiens* (Paris, 1789).
Anna Raitière, *L'Art de l'acteur selon Dorat et Samson* (Geneva, 1969).
Sieur [P.] Rameau, *Le Maître à danser, qui enseigne la manière de faire tous les différents pas de danse dans toute la régularité de l'art, et de conduire les bras à chaque pas* (Paris, 1726).
[René Rapin,] *Réflexions sur l'usage de l'éloquence de ce temps* (Paris, 1671).
Règles de la bonne et solide prédication (Paris, 1701).
Junius-Julius Regnault-Warin, *Mémoires de Talma* (Paris, 1804).
Marcel Reinhard, *Nouvelle Histoire de Paris: La Révolution, 1789–1799* (Paris, 1971).
[Toussaint de Rémond de Saint-Mard,] *Réflexions sur l'opéra* (The Hague, 1741).
Pierre Rémond de Sainte-Albine, *Le Comédien* (Paris, 1747).
Remontrances de MM. les Comédiens Français au Roi pour obtenir de Sa Majesté la suppression d'un arrêt du Conseil qui leur défend les ballets sous peine de 10,000 livres d'amende (n.p., 1753).
Jules Renouvier, *Histoire de l'art pendant la Révolution* (Paris, 1863).
Réponse des auteurs dramatiques soussignés à la pétition présentée à l'Assemblée nationale par les directeurs de spectacle (Paris, 1791).
François Riccoboni, *L'Art du théâtre*, (Paris, 1750).
Louis [Luigi] Riccoboni, *Pensées sur la déclamation* (Paris, 1738).
Sybil Rosenfeld, *Temples of Thespis: Some Private Theatricals in England and Wales, 1700–1820* (London, 1978).
Rousseau, *Lettre à M. *** sur les spectacles des boulevards* (Brussels, 1781).
Jean-Jacques Rousseau, *Considérations sur le gouvernement de Pologne*, in *Œuvres complètes*, ed. Bernard Gagnebin and Marcel Raymond, 4 vols. (Paris, 1959–69), III.
—— *Dictionnaire de musique* (Paris, 1768).
—— *Émile ou De l'éducation*, ed. François and Pierre Richard (Paris, 1964).
—— *Lettre à M. d'Alembert sur les spectacles*, ed. M. Fuchs (Lille and Geneva, 1948).

―― *La Nouvelle Héloïse*, ed. René Pomeau (Paris, 1960).
Gabriel de Saint-Aubin, Prints and Drawings (Baltimore, 1975).
Rémy G. Saisselin, 'Ut pictura poesis: Du Bos to Diderot', *Journal of Aesthetics and Art Criticism*, XX (1961-2).
Louis de Sanlecque, *Poème sur les manuvais gestes*, in Dinouart, *L'Éloquence du corps*.
Aldo Scaglione, *The Classical Theory of Composition* (Chapel Hill, 1972).
Antoine Schnapper, ' "Le Chef d'œuvre d'un muet", ou La Tentative de Charles Coypel', *Revue du Louvre et des musées de France*, XVIII (1968).
[Dom Sensaric,] *L'Art de peindre à l'esprit*, 3 vols. (Paris, 1758).
Jean Seznec, *Essais sur Diderot et l'antiquité* (Oxford, 1957).
Albert Soubies, *Les Membres de l'Académie des beaux-arts depuis la fondation de l'Institut*, 4 vols. (Paris, 1904-11).
Les Spectacles de Paris, ou Calendrier historique et chronologique des théâtres (Paris, 1792).
Les Spectacles des foires et des boulevards de Paris (Paris, 1777).
John R. Spencer, 'Ut rhetorica pictura', *Journal of the Warburg and Courtauld Institutes*, 20 (1957).
Anne-Louise-Germaine Necker, baronne de Staël, *Considérations sur la Révolution française*, 3 vols. (Paris, 1818).
Hippolyte Taine, *Les Origines de la France contemporaine*, 6 vols. (Paris, 1876-94).
François Talma, 'Réflexions sur Lekain et sur l'art théâtral', in *Mémoires de Lekain* (see above).
Tarare au Salon de peinture, 2 vols. (*Collection Deloynes*, XV, nos. 376 and 377).
Élisabeth Tardif, *La Fête* (Paris, 1971).
S. S. B. Taylor, 'Le Geste chez les 'maîtres' italiens de Molière', *XVIIe Siecle*, 132 (1981).
Jacques Thibault, *Les Aventures du corps dans la pédagogie française* (Paris, 1977).
Jacques Thuillier, 'Temps et tableau: la théorie des "péripéties" ', *Stil und Überlieferung in der Kunst des Abendlandes*, 3 vols. (Berlin, 1967), III.
B. R. Tilghman, *The Expression of Emotion in the Visual Arts. A Philosophical Enquiry* (The Hague, 1970).
'Timon', *Étude sur les orateurs parlementaires*, 2nd edition (Paris, 1837).
Robert Tomlinson, *La Fête galante: Watteau et Marivaux* (Geneva, 1981).
Georges Touchard-Lafosse, *Histoire parlementaire et vie intime de Vergniaud, chef des Girondins* (Paris, 1848).
[Tournon,] *L'Art du comédien vu dans ses principes* (Amsterdam and Paris, 1782).
Lionel Trilling, *Sincerity and Authenticity* (London, 1972).
Wesley Trimpi, 'The Meaning of Horace's Ut pictura poesis', *Journal of the Warburg and Courtauld Institutes*, 36 (1973).
Nicolas-Charles-Joseph Trublet, *Panégyriques des saints, suivis de réflexions sur l'éloquence en général et sur celle de la chaire en particulier*, 2nd edition, 2 vols. (Paris, 1764).

Graeme Tytler, *Physiognomy in the European Novel: Faces and Fortunes* (Princeton, 1982).
Jacques Ulmann, *De la gymnastique aux sports modernes* (Paris, 1965).
Charles-Augustin Vandermonde, *Essai sur la manière de perfectionner l'espèce humaine*, 2 vols. (Paris, 1756).
Jean Verdier, *Cours d'éducation à l'usage des élèves destinés aux premières professions et aux grands emplois de l'état* (Paris, 1777).
Vérités agréables ou Le Salon vu en beau par l'auteur du Coup de patte (Paris, 1789).
Paule-Monique Vernes, *La Ville, la fête, la démocratie: Rousseau et les illusions de la communauté* (Paris, 1978).
Georges Vigarello, *Le Corps redressé* (Paris, 1978).
Claude Villaret, *Considérations sur l'art du théâtre dédiées à M. Jean-Jacques Rousseau, citoyen de Genève* (Geneva, 1759).
Villiers, *L'Art de prêcher*, in Dinouart, *L'Éloquence du corps* (see above).
André Villiers, *L'Art du comédien* (Paris, 1959).
Roland Virolle, 'Noverre, Garrick, Diderot: pantomime et littérature', *Motifs et figures* (Paris, 1974).
François-Marie Arouet de Voltaire, *Lettres philosophiques*, ed. F. A. Taylor, 2nd edition (Oxford, 1946).
―― *Le Mondain*, in *Œuvres complètes*, ed. L. Moland, 52 vols. (Paris, 1877–85), X.
A. Vulpian and Gautier, *Code des théâtres* (Paris, 1829).
Daniel and Georges Wildenstein, *Louis David: Recueil de documents complémentaires au catalogue complet de l'œuvre de l'artiste* (Paris, 1973).
Georges Wildenstein, 'Talma et les peintres', *Gazette des beaux-arts*, 55 (1960).
Marion Hannah Winter, *The Pre-Romantic Ballet* (London, 1974).
Antoine Yart, *Mémoires ecclésiastiques et patriotiques, où l'on démontre la nécessité de transférer les fêtes au dimanche* ('Philadelphia', 1765).
Arthur Young, *Travels in France during the Years 1787, 1788 and 1789*, ed. M. Betham-Edwards (London, 1889).
Theodore Ziolkowski, 'Language and Mimetic Action in Lessing's *Miss Sara Sampson*', *The Germanic Review*, XL (1965).
Pierre Zoberman, 'Voir, savoir, parler: la rhétorique et la vision', *XVIIe Siècle*, 133 (1981).

Index

academicism 26, 47, 149, 150, 151–2, 158–9
Académie de Saint-Luc 151
Académie française 173
Académie royale de danse 117, 141, 150
Académie royale de musique 15, 68, 98, 100–1, 115, 117, 141
Académie royale de peinture et de sculpture 9, 30, 47, 76, 85, 150, 151–2
academy 15, 26, 149–51, 163, 173
 of acting 15, 139, 141, 143, 152, 159, 166, 172
acting:
 Elizabethan 65
 morality of 2, 17
 recording of 14, 16, 143–4, 146–7
 seventeenth-century French 66
 status of 1–2, 12, 18, 72, 141, 165–6, 169–70
 teaching of 2, 15, 64, 70–2, 125, 139, 166, 171–2
 theory of 9, 12, 49, 137, 151, 155, 157, 162, 165–6, 172
 and rhetoric 1, 10, 12, 18, 72, 152–3, 165, 169, 170–2
 actio 1–2, 10–11, 17–21, 25–6, 28–9, 31, 33, 35, 40, 44, 46–7, 48–52 *passim*, 61–6, 69, 71–3, 77, 81, 83, 87, 93, 109, 111, 122, 142–4, 146, 148, 155, 157, 159–61, 163, 165, 170, 175–6
 see also eloquence, bodily description of 2, 146
action ballet 6, 15, 25–6, 112, 126, 128–30, 135–7
actor:
 detachment of from role 30–1, 140

education of 123–4, 141–2, 152, 164–5, 170
involvement of in role 31
moral character of 166–9, 174
status of 1–2, 10, 15, 17, 24, 26–7, 72, 151–2, 159, 164–5, 168, 172–6
 see also acting, status of
Adry 85n.
Alard 95, 100–1, 103, 108
Alberti 11–12, 152
Ambigu-Comique 95, 98, 105, 107
Ambroise 105
Anecdotes curieuses et peu connues sur différents personnages qui ont joué un rôle dans la Révolution 56
Année littéraire 84, 156–7
Apologie du goût relativement à l'opéra 127
Argenson, d' 100–1
Aristippe 75, 170, 172–3
Aristotle 16, 18, 30–1, 36, 46, 48, 113, 125, 155, 170
Arnauld 52
Arnould, Sophie 69, 127
Arnould-Mussot 105
ars (versus *ingenium*) 151, 160
artist, character of 166–70, 175
assembly, national 19–20, 56–9, 168
attitude, physical 1, 7, 9–10, 12–13, 26, 63, 65, 70–1, 73, 75–6, 81–2, 89, 93, 125–8, 130, 139, 142, 144, 149–50, 156–7, 168, 175
Aubignac, d' 69
Aubry 87
Aude 168
Audinot 76, 95, 98, 102, 104, 106–8, 110, 113
Aumet 120
Austin 14, 16, 67, 69, 140, 143–4

Bachaumont 79, 102, 104, 107–8, 127, 135–6
ball 116, 119
ballet 98, 108 113–14, 116–17, 122–3, 125–8, 133–4
see also action ballet
Barère 59
Baron (writer on dance) 113, 119, 124–5, 132–3
Baron, Michel 69, 141, 146
Bary 50–2
Bathyllus 112
Batteux 18
Baume, Dame de 98
Beauchamps 114, 124–6
Beaujolais, troupe of comte de 102, 105
Beaumarchais 77, 89
Bellecourt 76
Berthélemy 82
Bertinazzi 77
Biancoletti 86
Bilcoq 79
Bissy, de 140
blind, the 41
Boïeldieu 55
Boileau 17, 37
Boissy d'Anglas 21
Bonnet 99, 125
Bossuet 2, 17, 29, 37, 45, 54–5
Boucher 81, 91, 168–9
Boulenger de Rivery 99
boulevard theatre 28, 95, 102, 104–5, 107, 113, 120
Bourdaloue 54–5
Bourdelot 99, 125
Boze 61
Brenet 82, 157
Brizard 76
Bulwer 64–5
Burke 20, 38–9

Cahusac 113, 125, 130, 132, 137
Caignez 172
Cailhava 66, 90–1
Callot 53
Campra 121
Carracci, Annibale 34, 75

Castiglione 161, 164
catharsis 31–2
Cato 166, 169
Caylus 81, 86, 145
Cerutti 54
Challe 82
Chaptal 174
Chardin 41, 90, 151
Charpentier 114
Chénier, Marie-Joseph 25n., 34, 60
choreography 16, 124–5, 135–6, 149, 175
Cicero 11–12, 37, 48–9, 64, 118, 158, 166, 169–70
civility 18–19, 125, 130, 157, 161–4
Clairon, Mlle 15, 64, 67, 72, 74, 76, 91–2, 125, 140–1, 144, 154, 156
Cléry 106
Clodion 82
Cochin 86
Colisée 96, 110
Collé 132
Collège des Quatre Nations 174
Collot d'Herbois 56
Colon 102
Comédie-Française 4, 6, 8, 15, 24, 46, 50, 58, 83, 98, 102–3, 115–16, 119, 123, 129, 139–40, 145, 166, 172–4
Comédie-Italienne 6, 8, 19, 98, 102–3, 145
Comédiens Français 4, 74, 86, 100–1, 103, 111, 115, 166, 174
Comédiens Italiens 86
Commedia dell'arte 7–8, 28, 86, 145, 148–9
Compan 117, 119, 121, 126–7, 129, 131–2, 137
composition 1, 11, 88–9, 133
'composition of place' 29
Condé 54
connoisseur 29, 78
Conservatoire 172
'constants' 157
Contat, Mlle 15
control, in performer 31, 66–72, 120, 128, 158–9, 163, 171

Index

convention 19, 35, 41, 74, 84, 122, 126, 129–30, 153, 156
  in drama 6–7, 156
Corneille 6, 24, 43, 76–7, 85, 115, 136, 146–7
Correspondance littéraire 73, 82, 106
correspondence of arts 18–19, 71, 152, 156–7, 159, 171
  see also painting and poetry; visual arts and literature
Corse 76
costume 74–5
coup de théâtre 90, 97, 131–2
Coypel, Antoine 13–14
Coypel, Charles-Antoine 12–14, 91
Cupid 91

Dallainville 173
dance 15–16, 19, 25, 80–2, 84–5, 99, 112–38 *passim*, 172, 175–6
  prestige of 1, 15, 117
  recording of 16, 124–5
  teaching of 1, 112, 125, 137–8, 172
Danchet 121
Dancourt 76, 86
David 9, 23, 66, 151–2, 159
  Brutus 15, 76, 85
  Les Sabines 77
  Le Serment des Horaces 15, 84–5, 93, 131
  Le Serment du Jeu de Paume 14, 39
  Socrate au moment de boire la ciguë 15, 85, 87
deaf, the 64, 73
declamation 15–16, 27, 48–9, 57, 60, 64, 66, 72, 75, 108, 123, 128, 139, 141, 158–60, 170, 172, 174
decorum 83, 156–9, 161, 163, 170
De l'organisation des spectacles de Paris 58
Demosthenes 37
député 19–20, 56–9, 64
Descartes 2, 9, 36–7, 154
Desmoulins 60–2
Destouches 121
detachment, from artistic creation 32–3, 140, 160, 168

Dézallier d'Argenville 74
Diderot, *De la poésie dramatique* 80, 87, 112, 147–8
  Éléments de physiologie 40–1, 47
  Éloge de Richardson 89
  Entretiens sur 'Le Fils naturel' 2, 6, 40, 73, 77–8, 80, 83–4, 87–8, 108–9, 111–12, 117, 128–9, 134–5, 147, 158–9
  Essais sur la peinture 47, 73, 86, 150
  Le Fils naturel 2–4, 78, 87–9, 129, 133, 137, 160
  Lettre sur les sourds et muets 3–6, 18, 40, 45, 73, 80, 109, 159
  Le Neveu de Rameau 3, 67, 117, 130, 140, 154, 169, 175–6
  Paradoxe sur le comédien 3, 30, 33, 84, 140, 148, 171
  Pensées détachées sur la peinture 35, 153
  Le Père de famille 2–4, 78, 87, 89, 160
  Salons 5, 41, 47, 76–7, 80, 82–4, 90–1, 93, 149, 151, 153–4, 158, 169
  Supplément au Voyage de Bougainville 3
  and *actio* 2
  as art critic 26, 73, 76, 80–4, 90, 149, 153–4
  as playwright 2, 37, 40, 46, 63, 87, 147
  as theorist of drama 2, 4, 37, 83, 87, 89, 111–12, 129–31, 133, 137, 147–9, 160
Dieulafoy 77
Dillon 59
Dinouart 33, 52–3, 64–5
Dolet 101, 103
Dorat 68–9, 74–5, 109
Dorfeuille 70–2, 75–6, 84, 142, 155–6, 164–6, 170–1
Dorneval 103
drama:
  in Jesuit colleges 29, 73–4, 113
  morality of 17, 30, 53, 63, 167, 169

drama (cont.)
  performance of 2, 36, 44, 143, 156, 158, 161, 175
drame 3, 40, 63, 79, 83, 89, 112, 130, 133
Dubois-Goibaud 52
Dubos 2, 6, 8-10, 18, 28-9, 31-4, 42, 44, 75, 99, 111, 130, 142, 153
Dubroca, Louis 49, 65-6, 69-70, 142, 170
Duclos, Mlle 17
Ducoudray 100, 107
Dufresnoy 73
Dugazon 15
dumb, the 13, 64, 73, 152
Dumesnil, Mlle 67
Dumont 107-8
Dumont, Étienne 60, 62
Dumoulin 133
Dupré 127-8, 150
Durameau 82
Duras 15
Durazzo 77, 121
Du Roveray 60
Duval 173
Dying Gladiator 75

École des élèves protégés 85
Edmé 98
education 12, 118-19, 161-4
eloquence:
  bodily 1-4, 6-7, 10, 28, 34, 37, 41, 46, 111, 138, 140, 142-3, 146, 150, 158, 160-2
  see also actio
  verbal 3, 6, 140, 143, 162, 165
emotion 9, 11-12, 30-4, 51, 55, 69, 92, 112, 126, 130, 144, 153, 155
  see also passion
empathy 31-2, 160, 168
empiricism 28, 36
enargeia 16, 36
Encyclopédie 75, 112, 117, 124-5, 134, 149
Encyclopédie méthodique 113-14, 119, 123, 131, 133-4

Engel 16, 142-3, 154, 171
enthusiasm 31n., 151
Entretiens sur l'état actuel de l'Opéra de Paris 136
Erasmus 161-2
Essai sur l'opéra 123
États généraux *see* States-General
excommunication 174
exercise 163
  spiritual *see* 'spiritual exercise'
Exposition de peintures, sculptures et gravures du Salon (1759) 156
expression, facial 1, 7, 9-10, 12, 51, 53, 61, 65, 68-71, 73, 75, 92, 109, 126-8, 142, 146, 149, 153-4, 161-2

Fabre d'Églantine 56, 147
fairs 19, 28, 95, 108, 115-16, 144-5
Falconet 26
Faret 162
Favart 68, 77, 121
Ferrières, de 60n.
fête 20-4, 59-60, 96-7, 100, 118, 168, 174
'fête galante' 76, 145
Feuillet 124
figures, rhetorical 36
Fleury 15, 56, 58-9
foire Saint-Germain 6, 8, 98-9, 100, 103-4, 106
foire Saint-Laurent 8, 86, 98-100
forains 8, 97-8, 100-1, 103, 115
Fox 57, 59
Fragonard 5, 77
Francisque 103
Furetière 125
Fuzelier 103

Gachet 96, 107, 110-11, 154
Gardel 84, 120, 126
Garrick 57, 129, 143
genius 31-2, 140, 148, 150-1, 157, 169, 176
Geoffroy 67, 70, 120-1, 123, 126, 131, 159
Gérard 43-4
Gersaint 145

Gesamtkunstwerk 44, 108, 114n., 121, 159
gesture 1-3, 7, 10, 12-13, 25, 28, 45, 48, 50-1, 53, 58, 60-1, 64, 66-72, 80-1, 85, 87-8, 91-2, 94, 99, 101, 108-10, 112, 116, 122-3, 126-7, 129, 133, 139, 142-4, 146-7, 152-4, 161, 165, 168, 175
Gibert 36, 171
Gigogne, Dame 104
Gilles 82, 145
Gillot 7, 85-6
Girardin 61
Girondins 20, 62n.
Gluck 78
Graffigny, Mme de 104
Grandmesnil 26, 50, 173
Grands Danseurs du Roi *see* Théâtre de la Gaîté
graphic arts 41, 144-6
Gregory of Nazianzus 50
Greuze 41, 47, 77-9, 87, 153-5, 158-9
Grimarest 49
Grimm 5, 84, 134-5
Grobert 21
Guérin 148
Gueullette 106
Guide des amateurs à Paris 106
guild 151
Guillaume 125, 137
Guillemin 119, 124
Guyenet 100

Hallé 84, 154
Hamlet 148
Hannetaire, Servandoni d' 67-8, 96, 146, 152
Harlequin 7, 19, 45, 77
Heinel 131
Hérault de Séchelles 19, 57, 64
Herculaneum 132
Hercules 70, 75
Hesse, de 132
Hobbes 37
Hogarth 7-8, 45-6, 69, 143
Homer 85

Horace 18, 37, 99, 131, 157
Hôtel de Bourgogne 98
Huerne de la Mothe 57
hypotyposis 36

Idées sur l'opéra 123, 127
Ignatius 29
illusion 5, 33, 102, 104-5, 122, 175
image 20, 29-30, 34, 36-8, 46-7, 79, 91, 153
imagination 3-4, 35, 40-2, 86, 169
imitation 18, 33, 38, 86, 92, 99, 113-14, 128, 137, 150, 158, 169
ingenium (versus *ars*) 151, 160
inspiration 140, 148, 151-2, 160, 171
instinct, in artistic creation 70, 140, 151
Institut de France 26, 50, 159, 173-4
inventio 155
Italians, acting of 1, 7, 85

Jason 92-3, 156
Jaucourt, de 149
Javilliers 133
Jeaurat 78
Jesuits 29, 54, 73, 113-14, 117, 161
Joli 68
Journal de Paris 4, 39, 42-4, 78, 81
Journal des théâtres, de la littérature et des arts 75
Journal encyclopédique 80-1
Jouy 77
Juvenal 99

Kemble 143

Lablée 96-7
La Bruyère 162
Lacouture 85n.
Lacretelle 61n.
Lafayette, Mme de 162
La Fontaine 5
La Fosse 77
Lagrenée 41, 81-2
Lairesse 35-6

La Motte 5, 121
Lamy, Bernard 36
Lamy, François 36
Lang 73–4
Laocoön 75
la Place, de 76, 101, 103
La Porte, de 95, 97
La Revellière-Lépeaux 21–2
Larive 54–5, 70, 149
La Rochefoucauld 117, 162
La Salle, de 107n., 163
La Tour, de 167
L'Aulnaye, De 99, 108
Lebrun 2, 9, 12–13, 65–6, 81, 126, 149, 153–4
Lecouvreur, Adrienne 14, 69
Le Faucheur 50
Le Gras 48
Le Jay 113–14
Lekain 1, 6–7, 54, 61, 74, 76, 111, 139–41, 149, 152, 159, 168, 172
Lemercier 57
Lemercier, Népomucène 61
Lemierre 104
Le Moyne 81
Leonardo 12, 73, 152, 154
Lépicié 82
Lequinio 23
Lesage 98, 103
Lessing 18, 38, 42, 142
Le Sueur 66, 75, 81
Lesuire 82
Lettre critique à un ami sur les ouvrages de MM. de l'Académie exposés au Salon du Louvre 1759 92
Lettre critique sur notre danse théâtrale 113, 128, 131, 133
Lettre d'un artiste sur le tableau de Mlle Clairon 92
Lettre écrite à un ami sur les danseurs de corde et sur les pantomimes qui ont paru autrefois chez les Grecs et chez les Romains et à Paris en 1738 73, 99
Levacher de Charnois 69–70, 75
Levesque de la Ravaillère 16, 45
liberal art 1–2, 10, 12, 76, 139, 152, 165–6

Ligne, prince de 69, 159
Locke 6, 28, 37, 162–3
Lomazzo 152–3
Longchamp 77
Longepierre 92
Louis XIV 1, 15, 114, 117, 141
Louis XVI 59
Lucian 120, 124–5
Lully 116–17, 121
Lycée Républicain 172

Macbeth, Lady 109
Magne de Saint-Aubin 166–8
Mainbray 6
Maine, duchesse du 6, 104
Maintenon, Mme de 8
Mallet du Pan 20, 56
Malpied 118–19, 125
manneredness 80–1, 84, 149–50
Marcel 80, 149–50
Marie-Antoinette 96
Mariette 13
marionnette *see* puppet
Marmontel 34, 55, 57–8, 75–6
mask 7, 66, 126, 128
Massillon 54–5
Mathon de la Cour 153
Maurice 95, 100–1, 103
Maury 54
Medea 91–3, 156
Médicis, Marie de 131, 155
Mémoire sur les danses chinoises, d'après une tradition manuscrite de quelques ouvrages de Confucius 118
memory 41, 168, 170
Ménageot 84, 156–7
Ménestrier 113, 117–18, 125, 129–30
Menus Plaisirs du Roi 15
Mercier 4, 42–4, 47, 56, 88–90, 94, 107, 132
Mercure de France 60, 75, 86
Mérilhou 61
Merlin de Thionville 22–3
'method' acting 110, 151–2, 171
Michelangelo 75
Millet 83
Millin de Grandmaison 168

Mirabeau 20, 58-62
Mirabeau à l'Assemblée constituante 61
Molé 15, 20, 59-60, 139, 142, 171-3
Molière 43, 66, 69, 90,114-15, 141, 146, 172
Mona Lisa 154
Moncrif 164
Mondory 69
Monet 98
Montesquieu 82
mountebank 144-5
movement, bodily 2, 7-8, 10, 25-6, 28, 45-7, 64-5, 68-9, 71, 80, 89, 92, 99, 109, 112, 126, 130-2, 137, 146-7, 150, 155, 157
Moy, de 23-4, 168
music 20, 46n., 143

Napoleon Bonaparte 27, 159
naturalness 66-72, 80-1, 84, 88, 127, 149, 157
Necker 57-9
neo-classicism 81, 93
Newton 14
Nicolet 83, 95, 100, 104, 107, 110, 173
Nolant de Fatouville 86
non finito 38, 47
Nougaret 108, 123
Noverre 6, 15, 24n., 25, 40, 85, 88, 97-8, 112, 117, 124-38 *passim*, 149, 160, 175

Observateur littéraire 79, 91
Observateur philosophique 78-9
Observations critiques sur les tableaux du Salon de l'année 1787 85
Observations d'une société d'amateurs sur les tableaux exposés cette année 1761 79
Observations sur l'exposition des peintures, sculptures et gravures du Salon du Louvre (*Observateur littéraire*) 91
Octavien 76
Odéon 166
Oldfield, Sara 14
opera 44-5, 68-9, 100, 112, 114, 116-17, 120-3, 131-6, 159
Opéra 6, 8, 15, 78, 81-4, 95, 98-100, 102-4, 114-15, 119-21, 135, 145, 174
opera buffa 116
Opéra-Comique 77, 98, 104, 140-1, 145
orator, ancient 10, 20, 32-3, 48, 53, 64, 72, 155, 165, 169
oratory 1, 10, 12
  ecclesiastical 1, 10, 34, 36, 49-55, 62-3, 68, 72, 161, 165, 174
  forensic 1, 10, 35, 37, 49-50, 62-4, 72, 161, 165, 174
  political 19-20, 48-9, 55-62, 67, 72, 161, 165, 174
Orléans, duc d' 100
ornament *see* ornateness
ornateness 36-7, 50, 63

painter, training of 74, 142
painting:
  as liberal art 1, 12
  genres of 79, 83, 87, 132, 151, 158-9
  morality of 35
  technique of 26, 90
  theory of 155-7
  and poetry 18, 37-9, 152, 155, 157, 159
  see also correspondence of arts; visual arts and literature
  and rhetoric 1, 11, 153, 155, 169
Palais-Royal 105
Panckoucke 113
Pantomimanie, La 94-5, 147
pantomime 1, 3, 6, 20-1, 25, 29, 37-8, 40, 43-4, 47, 66, 70, 79, 87, 94-111 *passim*, 113, 120, 124-30, 136, 147-8, 153, 158, 160
'papillotage' 88n., 133
Parfaict, François and Claude 97-8, 100, 103, 108
parlement 50, 57, 101, 103, 124
Pascal 117
passion 2, 8-9, 13, 16-17, 30, 33, 46, 55, 66, 91, 94, 99, 109-10,

passion (*cont.*)
  114, 117, 120–1, 124, 126–7, 130–1, 149, 152–5, 170
  see also emotion
pastoral 91
pathognomy 2, 9, 154–5
Pellissier, Mlle 104
perception *see* sensory perception; visual perception
Perico 105
peripateia 155
persuasion 2, 10–11, 28, 31–3, 35, 44, 49, 51, 54, 90, 165, 168, 170–1
Petits Spectacles de Paris, Les 102
Petronius 99
Philip of Neri 51
physiognomy 9, 70–1, 126, 128, 149, 154–5
physiology 33, 40, 47
Picard 173
pictorialism, verbal 34–8
Pierrot 7, 46, 76
Piles, de 73
Piron 103–4
Pixérécourt 172
Plato 4–5, 30, 35, 43, 46, 63, 85n., 88, 113, 117–18, 125, 169, 170
Plutarch 125, 129
Poetus and Arria 75, 148
Polichinelle *see* Pulchinello
Pompeii 132
popular entertainment 6, 8, 19–25, 52–3, 56–62 *passim*, 73, 76–7, 79, 82–3, 95–111 *passim*, 119, 128–9, 131, 144
Porte Saint-Martin 120–1, 123
Portrait de Mlle Clairon, par Carle Vanloo 92
position 125, 175
Poultier 56
Poussin 34, 66, 81
preaching *see* oratory, ecclesiastical
Premiers Gentilshommes de la Chambre du Roi 15n., 103, 140–1
Préville 7, 15, 26, 56, 59
Prévost 52

pronuntiatio 6, 34, 50
propriety 53, 62, 67, 71, 100, 109, 122, 156–8
  see also decorum
Pulchinello 7, 104
Punch 45
puppet 59, 102–5, 133, 148, 176
Pylades 112

Quatremère de Quincy 44–6, 116, 121–2, 136, 159
Quinault 116–17, 121, 131, 134
Quintilian 35–6, 48–9, 99, 118, 142

Racine 13, 31, 43, 63, 85–6, 115, 172
Rameau, Jean-Philippe 78
Rameau, Pierre 118–20, 124–5
Raphael 132
Rapin 49, 52
Raucourt, Mlle 140
recording *see* acting, recording of; dance, recording of
Regent *see* Orléans, duc d'
Règles de la bonne et solide prédication 50
Rembrandt 75
Rémond de Saint-Mard 114, 121, 125–6
Rémond de Sainte-Albine 44, 46–7, 152, 159
Reni 75
Renou 82
Revolution, French 14, 19–25 *passim*, 30, 34, 39, 54–6, 95, 102, 105, 111, 168, 173–4
Révolutions de France et de Brabant 60, 62
Reynolds 143
rhetoric:
  propriety of 36
  theory of 1, 10, 30, 37, 124, 157, 165–6, 170
  and ethics 32, 170–1
Riccoboni, François 6, 110, 152
Riccoboni, Louis [Luigi] 48–9, 66, 139

Riccoboni, Mme 90, 93
Richardson 89
Robin 157
Rollin 162
Roscius 158
Roslin 83
Rousseau 19, 88, 167, 169
  Considérations sur le gouvernement de Pologne 21
  Dictionnaire de musique 129, 134
  Discours sur les sciences et les arts 63
  Émile 55n., 161
  Lettre à M. d'Alembert sur les spectacles 22, 32, 63, 167
  La Nouvelle Héloïse 135–6
Roy 77
Rozetti, Mlle 68
Rubens 75, 131, 155
rules 15, 25, 64, 72, 119, 125–6, 139–40, 142, 149–51, 157–60, 171–2

Saint-Aubin 144–6
Sainte-Foy 85, 90
Salle des Machines 97
Sallé 95
Salon (du Louvre) 5, 9, 76–85 *passim*, 91–2
saltatio 6, 99, 112, 129
Sanlecque 54, 68
sans-culottes 75n.
Scaramouche 7
Sceaux 6, 104
Second 104
Sensaric 35, 37
sensationalism 6, 25, 34
sense-impression 10
sensibility 31, 70, 119, 126, 133, 160, 167–8
sensory perception 4–6, 21, 28, 46, 79
  see also visual perception
Séraphin 106
Servandoni 96–7
shadow-puppet 105–7
Shaftesbury 18, 42, 155
Siddons, Mrs 143
sight 6, 21, 28, 34–5, 41, 79

sincerity 33, 51, 53, 63
sketch 38, 43, 144–6
Socrates 170
Spectacles de Paris, ou Calendrier historique et chronologique des théâtres, Les 103
Spectacles des foires et des boulevards de Paris, Les 107
speech, suppression of 8
'spiritual exercise' 29
spontaneity 1, 8, 38, 67, 71, 110, 133, 140, 147–9, 151
States-General 59
Stendhal 9, 148
Steuben 148
'still' 144, 148
Sulla 148
Supplément du peintre anglais 84n.
Suvée 82

tableau 6, 34, 77, 87–91, 93, 97, 108–9, 122, 131–4, 144
Taine 57
Talma, François-Joseph 20, 24, 27, 53–4n., 62, 67, 70, 74, 76, 141, 148–9, 152, 159, 171–2
Talma, Julie 20, 62n.
Tarare au Salon de peinture 83
Tasso 85
Télémaque 98
Teniers 132
Terence 99
Théâtre de la Gaîté 95, 100, 107
Théâtre de la Nation *see* Comédie-Française
Théâtre de l'Égalité 174
Théâtre de la République 25, 50, 58
Théâtre des Associés 95
Théâtre des Délassements Comiques 95, 102
Théâtre des Marionettes 103
Théâtre du Lycée Dramatique 95
Théâtre du Vaudeville 77
Théâtre-Français *see* Comédie-Française
Théâtre-Italien *see* Comédie-Italienne
Théâtre Saint-Martin 120, 123

Théâtre sans Prétention 94
Théâtre Séraphin 105–6
'theatricality' 12, 49–55 *passim*, 61, 72, 80–1, 83, 88, 90–1, 93, 156, 174
'tiers état' 59
Timanthes 75
Titian 34
Torré 96
Touchard-Lafosse 61
Tournon 141
training schools 25, 139
  see also academy; acting, teaching of; actor, education of; painter, training of
Troyes, de 81
Trublet 54
Turenne 173

Valcour 102
Vandermonde 163
Vanloo, Carle 76–7, 80–1, 91–3, 144, 149, 156
Vaucanson 149
Vaux-Hall 96
Verdier 163
Vergniaud 61
Vestris 84, 126–7, 131, 136, 149

Vien 76
Vigée-Lebrun, Mme 105
Villaret 167
Villeterque 43–4
Virgil 85
visual arts 9, 12, 29, 41, 73–93 *passim*, 142–3, 154, 156, 174
visual arts and literature 18, 38, 42–4, 159
  see also correspondence of arts; painting and poetry
visual perception 4–5, 28–30, 40, 46–7, 79
  see also sensory perception
vividness *see enargeia*
Voltaire 6–7, 14–15, 24, 34, 43, 76–7, 85, 104, 111, 121, 147

Watteau 7, 74, 76, 85–7, 144–5, 154–5
Weaver 6
Wille 79, 83
word, expressiveness of 2–4, 10–11, 34, 37–8, 42, 46–7, 147–8
Wright 65

Young 58